Community ecology and salamander guilds

CAMBRIDGE STUDIES IN ECOLOGY

ALSO IN THE SERIES

Andrew J. Beattie *The evolutionary ecology of ant–plant mutualisms*
Jeremy J. Burdon *Diseases and plant population biology*
Hugh G. Gauch, Jr. *Multivariate analysis in community ecology*
Kenneth A. Kershaw *Physiological ecology of lichens*
Robert P. McIntosh *The background of ecology*
Robert H. Peters *The ecological implications of body size*
C. S. Reynolds *The ecology of freshwater phytoplankton*
F. I. Woodward *Climate and plant distribution*

Community ecology and salamander guilds

NELSON G. HAIRSTON, SR.
Department of Biology
The University of North Carolina at Chapel Hill

The right of the University of Cambridge to print and sell all manner of books was granted by Henry VIII in 1534. The University has printed and published continuously since 1584.

CAMBRIDGE UNIVERSITY PRESS
Cambridge
New York New Rochelle Melbourne Sydney

Published by the Press Syndicate of the University of Cambridge
The Pitt Building, Trumpington Street, Cambridge CB2 1RP
32 East 57th Street, New York, NY 10022, USA
10 Stamford Road, Oakleigh, Melbourne 3166, Australia

First published 1987

Printed in the United States of America

Library of Congress Cataloging-in-Publication Data

Hairston, Nelson G.

Community ecology and salamander guilds / Nelson G. Hairston.

p. cm. – (Cambridge studies in ecology)

Bibliography: p.

Includes index.

ISBN 0 521 32578 1

1. Salamanders – Ecology. 2. Biotic communities. 3. Amphibians – Ecology. I. Title. II. Series.
QL668.C2H35 1987
597.6′5 – dc19 87–17767

British Library Cataloging-in-Publication applied for

For Patty

Contents

Preface

This book arose from two stimuli. The first was a desire to put the large amount of published work on salamander ecology into the context of other work on animal communities. Community ecology has a history that has been strongly influenced by the preferences of a few ecologists for particular kinds of animals. It is clear that MacArthur's fascination with birds remains a strong influence, as a perusal of the indexes of recent compendia on community ecology will show. A second group that has had a strong impact has been the lizards, but until recently, lizard ecologists followed or elaborated the theories of MacArthur. Thus, new work on birds or lizards tended to be incorporated into community theory, especially if it confirmed the prevalent models. Theoreticians paid little more than lip service to work in other areas, most specifically to the growing literature on experiments in marine environments. Specific examples demonstrating the existence of interspecific competition were cited, but there have been few attempts to incorporate this large amount of experimental ecology into community theory. Work on salamanders shared the fate of experiments in the rocky intertidal.

The second stimulus was an invitation from John Birks to write this book for the series Cambridge Studies in Ecology. The invitation came just when I had completed a very tiring experiment of four years' duration and had decided not to begin another effort at any such strenuous level. Thus, I was in a susceptible frame of mind.

The first chapter consists of an account, perhaps biased, of community ecology and the problems that are beginning to be faced by at least some workers in the field. It is followed by a comparison of the constancy of salamander populations and populations of groups that have contributed most importantly to community theory. Next, I felt that it was necessary to provide readers with an understanding of the variety of salamander life histories, before describing the research that has been carried out in the three conspicuously different habitats that have attracted the attention of salamander ecologists. It has been my experience in ecological discussions that relatively few biologists are aware that most salamander species do not follow the classical pattern of coming to ponds in early spring to lay eggs for embryologists. The differences among their life histories make a great deal of difference in the kinds of interaction that are to be expected in the different habitats.

It is rare that evolutionary questions can be attacked experimentally in the field, but unusual opportunities have been perceived and acted on in the case of salamanders. The conclusions from such work depend on the assumption that present ecological forces are the same as those that were involved in natural selection in the past, but if granted, the assumption allows important conclusions to be made. A final chapter describes the opportunities for further research on salamander communities.

Acknowledgments

I am grateful to many people whose cooperation and help have contributed to the successful completion of this book. John Birks proposed that I should undertake it and has been helpful and supportive throughout with his comments as the individual chapters have been completed. Kiisa Nishikawa, Alan Stiven, and Nelson Hairston, Jr., read and made valuable comments on various parts. Steve Tilley read the whole manuscript carefully and made a number of perceptive comments. As much as I would like to do so, I cannot blame any of them for such errors as may remain. The many authors and publishers who graciously gave permission for me to copy figures and tables are acknowledged in the appropriate places, but I hereby thank them profusely.

My experiments, described in Chapters 5 and 6, could not have been completed without the help, most of it volunteered, of family members, colleagues, students, and friends. Clambering around steep mountainsides at night, sometimes in the rain, is exhausting work. The following have earned my lasting gratitude: Steve Arnold, Bill Bowen, Bill Bramble, Judy Edson Bramble, Ellen Gilinsky, Tim Goater, Nelson Hairston, Jr., Patty Hairston, Lynn Houck, Ken Kneidel, Bruce Means, Kiisa Nishikawa, Ron Nussbaum, Susie O'Bryan, Ray Plotecia, Julie Roberts, Bill Searcy, Margaret Searcy, Phil Service, Chuck Smith, Mark Southerland, Sally Stenhouse, Steve Tilley, Martha Weston, JoAnn White, Ann Williams, and Del Williams.

1

Challenges to community theory

Introduction

Community ecology is a subject in ferment. After a decade or more of virtually universal acceptance, community theory based on interspecific competition and niche partitioning is under challenge from several directions. The so-called null models have received the most attention and have drawn the most vigorous responses from proponents of the conventional theories of community evolution and their biogeographic corollaries. Null models are not the oldest challenge to community theory. All forms of community theory depend on the importance of interspecific interactions, and these in turn cannot be important without a negative relationship between the density of a population and the rate of increase in its numbers. Now, whereas the promulgators of null models argue that interspecific interactions, especially the kind used commonly in community theory, remain to be demonstrated, their arguments are quite different from those of the proponents of density-independent factors as the dominant determinants of the abundance of organisms. One could say that those who argue in favor of density-independent factors would automatically support null models, but that proponents of null models do not necessarily agree with the density-independent school, because they do not deny the importance of interspecific interactions.

In this chapter, I describe a number of models in community ecology, with some of their evolutionary correlates. Mathematical models have been used extensively in ecology. They have the great merit of requiring both the creator and the user to think clearly and to be precise as to the meaning of the implications of the terms used. It is important to understand the assumptions that were made in constructing the models. The challenges in the title of the chapter come from pointing out unsatisfactory assumptions, from field experiments, the results of which refute the validity of one or more models, and from arguments that confirmation of their validity is based on poor application of the scientific method. The criticisms are offered in a constructive spirit and in the hope that theoretical community ecology of the future will take into account the results of the large number of experiments and will be such that specific experimental tests of the principles can be devised.

It is beyond the scope of this book to cover the field completely, and attention is centered on those aspects of community theory that lend themselves to testing experimentally. Salamander communities have been shown to provide excellent material for these tests, and experiments involving them will be described in detail in later chapters. More complete mathematical statements of community theory can be found in Roughgarden (1979), May (1981a), Vandermeer (1981a), and Pimm (1982).

Descriptive models

The first quantitative models of natural communities were mathematical descriptions of the relative abundances of species (Fisher, Corbet, & Williams 1943; Preston 1948, 1962a,b). These models proposed no mechanisms to account for the observed frequency distributions, and for some communities it was shown that the constants varied with the inclusiveness of a sample from the community (Hairston & Byers 1954). These patterns continue to attract some attention (Sugihara 1980; May 1981b), but since the appearance of truly mechanistic models, most ecologists have ignored them.

MacArthur (1957, 1960) proposed the first models that started from biological premises. In his "broken-stick" models of animal communities, he suggested that a limiting resource could be thought of as being divided up among species in several different ways, and he proposed that a choice among the models could be made by comparing them with observed species abundance distributions. The approach attracted much attention (e.g., King 1964), but some predictions were found not to be upheld (Hairston 1959), and the mathematical development of one of the models was incorrect (Pielou & Arnason 1966; Vandermeer & MacArthur 1966). This caused MacArthur (1966) to abandon the whole approach, although it continues to appeal to some ecologists (e.g., Bush & Holmes 1983).

The niche-preemption model, analogous to the broken-stick models (Whittaker 1965), involves similar biological premises, but different mathematical relationships.

Mechanistic models

The development of current theories of community composition began with the papers of MacArthur and Levins (1964, 1967). The approach was through the familiar Lotka–Volterra equations describing competition between two species. The equations are themselves extensions of the logistic equation for population growth. As with any mathematical model of a biological process, the equations contain certain assumptions. Some of the assumptions are biological; others are within the mathematics. If the mathematical manipulations involved in the construction of a model have been properly carried out, the conclusions reached from the application of the model are dependent on the assumptions. It is therefore essential that the assumptions be identified, and evidence of their validity be produced, before the conclusions can be accepted.

I show in later chapters that salamanders meet some of the important assumptions better than most other kinds of organisms, and then I describe the use of these animals in testing models of community organization, including well-known null models.

The assumptions of the logistic model of population growth have been discussed in detail by Smith (1952), Slobodkin (1953, 1962), and Andrewartha and Birch (1954). They are carried over into the Lotka–Volterra equations for interspecific competition and into the community models, except where the latter have been explicitly modified to alter particular assumptions. It is well to recall them. The equation is

$$dN/dt = rN(K - N)/K \qquad (1.1)$$

where r is the intrinsic rate of natural increase, N is the number of individuals present, and K is the maximum number that can be supported by the specified environment. The model assumes that all individuals have the same impact on the population growth rate per individual, $dN/N\,dt$, regardless of age, sex, or population density. The growth rate per individual declines in linear manner from rate r when N approaches zero to a rate of zero when N reaches K. The equation is valid only when the population starts at the stable age distribution: that condition in which all age classes are increasing at the same rate, and in which each age class is represented in the same proportion at successive times. If K is determined by a consumable, renewable resource, that resource is assumed to be renewed at a constant rate. If it is set by some fixed resource, such as space, the assumption of equality of effects by different individuals is unlikely to be correct for different densities.

Every ecologist can provide objections to each of the assumptions, but the equation and the models using it as a starting point continue to be used, sometimes with claims of validation in various observational and experimental situations. A widely quoted example is the yeast culture to which Pearl (1925, p. 10) applied the logistic equation, despite Smith's analysis (1952) showing that several of the assumptions were not met in the culture. In the general enthusiasm for elegant mathematical models – called "physics envy" by Cohen (1971) – these important and largely unrealistic assumptions have been overlooked, ignored, or glossed over by a large number of ecologists.

The Lotka–Volterra model of interspecific competition considers two or more species by modifying the logistic equation with terms that allow individuals of all other species to have effects on the population growth rate for each species in the competing group:

$$dN_i/dt = r_iN_i(K_i - N_i - \sum\alpha_{ij}N_j)/K_i \tag{1.2}$$

where α_{ij} (the competition coefficient) is the effect of an individual of species j on the population growth rate of species i, relative to the effect of an individual of species i on its own population growth rate, and the other terms are as in the logistic equation. The competition coefficient is constant for all densities of all species, and the effect of adding an individual of any of the species is instantaneous; that is, time lags are assumed to be negligible. The analyses begin with the further assumption that the populations are in equilibrium, and therefore dN_i/dt is zero. This assumption collapses the equation to

$$K_i = N_i + \sum\alpha_{ij}N_j \tag{1.3}$$

and r_i does not need to be estimated in order to specify the relationships numerically. If either K_i or α_{ij} can be estimated, the other can be found.

Plainly, the most direct method would be to remove members of species j and observe an increase in the number of species i. The ratio of the number of i added to the number of j removed is α_{ij}. This approach measures the effect of one species on the other. It does not detect the nature of the interaction, whether it be competition for some consumable resource, competition for a fixed resource, or interference with the ability of species i to obtain a critical resource, or, in some cases, whether the interaction is competition or predation (Hairston 1980a, 1986). Moreover, it is "labor-intensive, time-consuming and expensive" (Rosenzweig et al. 1985, p. 194).

Not surprisingly, many ecologists have turned to indirect methods to estimate the coefficients (MacArthur & Levins 1967; Schoener 1974b; Hallett & Pimm 1979). MacArthur and Levins showed mathematically the conditions necessary for a third species to invade a habitat occupied by two species, when all three are dependent on overlapping parts of a single resource. The conditions concern the amount by which the demands of each species overlap the demands of its competitor(s), taking into consideration the variance in use of a range of the resource by each. This "limiting similarity," then, will determine the number of species that can coexist and hence will determine the structure of the community. It depends heavily on the assumption that overlap is *the* measure of interspecific competition, and hence represents α_{ij}. The original MacArthur–Levins approach was to estimate the overlapping proportion of use of the assumed limiting resource by each of two species. The product of the overlapping proportions, divided by the square of the proportion of the first species' use of the resource, was equated with the competition coefficient:

$$\alpha_{ij} = \sum p_{ih} p_{jh} / \sum p_{ih}{}^2 \tag{1.4}$$

Schoener (1974b) considered the necessity of taking into account the renewable nature of many resources and the fact that their equilibrium densities in the environment may be different. He modified the original appealingly simple formula accordingly:

$$\alpha_{ij} = \left(\frac{T_j}{T_i}\right) \left[\frac{\sum_{k=1}^{m} (d_{ik}/f_k)\,(d_{jk}/f_k)\,(b_{ik}K_k/r_k)}{\sum_{k=1}^{m} (d_{ik}/f_k)^2\,(b_{ik}K_k/r_k)} \right] \tag{1.5}$$

T_i is the total number of resource items utilized per unit time by competitor i; d_{ik} is the frequency of resource k in the diet of i; f_k is the frequency of resource k in the environment; b_k is the net energy per item of resource k extractable by an individual of competitor i; r_k is the intrinsic rate of increase of resource k; K_k is the carrying capacity of the environment for resource k in the absence of predation; and m is the total number of different resources. Schoener suggested some simplifying reductions, but admitted the formidable practical problems. Indeed, one could characterize any attempt to compute the competition coefficients as labor-intensive, time-consuming, and expensive. A notable example is that of Spiller (1986), who obtained values for all of the terms except for the carrying capacities and intrinsic rates of increase of the resource items. His calculated values of the competition coefficients were "qualitatively consistent" with experimentally determined effects of competition on spider food-consumption rates.

Hallett and Pimm (1979), following Schoener (1974c), proposed to estimate the coefficients of competition between two species by comparing their respective densities in a large number of habitats. The data would then be used by assuming the equilibrium condition for the two-species Lotka–Volterra equation

$$N_{1h} = K_1 - \alpha_{12} N_{2h} \tag{1.6}$$

as a regression equation with N_{1h} as the dependent variable, and K and α as the regression coefficients. In order to allow for differential habitat selection by the two species, they proposed a more elaborate regression using the components of the different habitats as additional partial regression coefficients. This method obviously carries all of the assumptions in the Lotka–Volterra equations. Moreover, as Schoener recognized, it is assumed that the competition coefficients are constant over all of the habitats. There is also the practical problem of making a sufficient number of density estimates in different habitats, and assessing independently the effect of each habitat on the selection by each species.

There have been at least two tests of the validity of the Hallett–Pimm method of estimating competition coefficients. The first can be described as inadvertent. Hallett (1982) used the method on guilds of desert rodents. He found statistically significant negative coefficients within the *Perognathus–Peromyscus* guild, but not within the *Dipodomys* guild, and he also found no significant negative coefficients between these two guilds. Meanwhile, a different approach to competition in this community had recently been undertaken by Munger and Brown (1981), apparently too late to be included in Hallett's discussion. They constructed fences around four plots each 50 × 50 m. In these fences, they made holes 1.9 cm in diameter – large enough to allow passage of small granivorous and small omnivorous rodents. The granivores present included two species that were investigated by Hallett (*Perognathus penicillatus* and *Peromyscus maniculatus*), plus an additional species of *Perognathus* and *Peromyscus eremicus* (which they classified as an omnivore, though it had been placed in the same guild as the others by Hallett). The fence excluded the three species of *Dipodomys*, which were removed by trapping. Two of the species of *Dipodomys* were the ones investigated by Hallett. Control plots were also fenced, but the holes were 6.5 cm in diameter – large enough to allow passage of *Dipodomys*. The results were unequivocal. Exclusion of *Dipodomys* led to a statistically significant increase in the abundance of the small granivores, but not of the omnivores. Thus, the results of the direct experiment contradicted the indirect regression method.

The second test of the Hallett–Pimm method was intentional. Rosenzweig and associates (1985) trapped desert rodents in seven sites in arid parts of Israel. After exhaustive statistical analysis of their data, they concluded that the results were too inconsistent to be relied on for estimation of coefficients of competition. The paper has aroused a controversy, with Pimm (1985) and Schoener (1985a) arguing that the consistency among methods is better than admitted by Rosenzweig and associates. It seems unlikely that the matter will be settled without satisfactory simultaneous experiments and regression analyses on the same species in the same areas.

The use of any of these indirect methods requires acceptance of the assumption that competition is taking place with no more evidence than common occurrence and the use of some resource in common. Because the resource can be habitat, common occurrence plus the suspicion that the species might be competing are the only requirements. The methods are popular for the very reason that the assumption gives the appearance of relieving the investigator of the necessity of demonstrating that the species are competing. Attempts to validate the assumption are not always successful, especially when done in a noncircular way, as by a field experiment. The most convincing experiments on niche partitioning as a result of competition have been those of Werner (1977) and Werner and Hall (1976, 1977, 1979). In a series of experiments with three species of sunfish, they showed that interspecific crowding resulted in a shift of *Lepomis macrochirus* from feeding in the pond-edge vegetation to feeding in open water, and of *L. gibbosus* from feeding in the vegetation to feeding on bottom-living prey. *Lepomis cyanellus* continued to feed in the vegetation. It was shown to be a better competitor there than *L. macrochirus*. Despite a lack of replication and problems of different densities in one experiment, this work represents the best experimental evidence for an effect of interspecific competition in causing niche partitioning.

Competition within guilds

The foregoing comments do not relate to experiments in which interspecific competition has been found. They have been the subjects of two recent reviews (Connell 1983; Schoener 1983). There is little doubt that interspecific competition is widespread in nature. The question being addressed here is the extent to which niche partitioning has been demonstrated experimentally.

Several ecological systems have been subjects of less successful experiments. It happens that five such attempts were published within less than two years; none reported confirmation of guildwide competition, and two failed to detect any competition in guilds where it had been expected.

Dunham (1980) carried out experimental removals of two species of lizards in Big Bend National Park, Texas. He obtained positive responses of each to the removal of the other, but only during the two dry years (out of four) of the experiment. He informed me that he observed no response by the other ten species of lizards in the area (Dunham, pers. commun.). Smith (1981) obtained similar results in an experiment on two species of lizards in Arizona. During the first year, which was unusually dry, he observed positive results of removing each of the species, but only on growth and survival of immature animals. There was no effect during the normal second year, nor did two rare species respond to the experimental removals. The experiment of Tinkle (1982) gave completely negative results. After nearly four years of close observation of the population histories of three species of lizards in central Arizona, he removed the most abundant and one other for a period of two more years. Based on its natural history in other areas, he expected a specific shift in habitat as a result of the removals. No shift occurred, and there were no other kinds of responses, although abundance, survival, and growth were followed closely. As in the other two experiments, the remaining species in the community showed no effect of the removal of the numerically dominant species.

Wise (1981) carried out removal experiments on darkling beetles (Tenebrionidae) over a period of three years in New Mexico. Five species were "common" in the community. One of them, *Eleodes obscurus*, accounted for more than 50 percent of the adult tenebrionid biomass in the community. Wise removed these adults from three fenced 0.1-ha plots and kept three other plots as controls. There was no increase in the abundance of adults of the other four species of Tenebrionidae during the three years, nor was there any increase in the mean sizes of those adults. He concluded that despite similar omnivorous food habits by the beetles, competition was unimportant in this community.

My test of the assumption of competition between related species was of a guild of seven species of terrestrial salamanders (Hairston 1981, 1984). I selected two areas in the southern Appalachian Mountains for the experiments. The original purpose (Hairston 1973) was to test for competition between the two most abundant species in that habitat (*Plethodon jordani* and *P. glutinosus*). Removal of either of them resulted in a positive response by the other, demonstrating that they were in competition, which was much stronger in the Great Smoky Mountains than in the Balsam Mountains (Hairston 1980a; see Chapter 5).

On the experimental plots there were five other species of Plethodontidae in the Great Smoky Mountains, and four other species in the Balsam Mountains. All of them spend most or all of their adult lives on the forest floor with the two competing species, eat the same kinds of food (insects and other small invertebrates), and use the same kinds of burrows as refuges. Thus, from the standpoint of theoretical community ecology, they qualify as members of the same guild as the competing species. The names and abundances on the control plots are shown in Table 1.1. Removal of either *P. jordani* or *P. glutinosus* had no statistically significant effect (Table 1.2). One of sixteen tests was "significant" at the 5 percent level – close to the proportion expected by random chance. It should be noted that one of these five species

Table 1.1. *Mean number of individuals per plot search*

	Great Smoky Mtns.		Balsam Mtns.	
	Control plots	All plots	Control plots	All plots
Plethodon jordani	27.48		35.36	
P. glutinosus	5.00		4.36	
P. serratus	0.69	0.54	1.12	1.05
Eurycea bislineata	0.80	1.09	0.20	0.18
Desmognathus ochrophaeus	3.19	3.25	0.17	0.25
D. wrighti	0.08	0.08	0.28	0.38

Note: Plethodon jordani and *P. glutinosus* were affected by the manipulations, and their numbers were not comparable to those of the other species, except for the control plots. "*Desmognathus ochrophaeus*" includes *D. imitator* in the Great Smoky Mountains.
Source: Hairston (1981).

Table 1.2. *Mean number seen per plot search*

	P. serratus	*D. wrighti*	*D. ochrophaeus*	*E. bislineata*
Great Smoky Mountains				
Controls	0.69	0.08	3.19	0.80
P. jordani out	0.67	0.07	4.75	1.27
P. glutinosus out	0.65	0.10	3.25	0.90
Balsam Mountains				
Controls	1.12	0.28	0.17	0.20
P. jordani out	1.48	0.40	0.05	0.32
P. glutinosus out	0.97	0.30	0.50[a]	0.08

Note: D. ochrophaeus includes *D. imitator* in the Great Smoky Mountains.
[a]Significantly greater than control ($p < .05$).
Source: Hairston (1981).

(*Plethodon serratus*) is congeneric with the two competing species, and reducing the "guild" to congeners would not solve the problem. It has been suggested (Schoener 1983) that the experiment was not conclusive, because competition between the less abundant species and the others could be of a kind that gave curved zero isoclines (described in a later section), which would require a greater reduction in numbers of *P. jordani* or *P. glutinosus* than was achieved (about 60 percent of the mean of controls). I showed (Hairston 1985) that in the case of *P. serratus*, it would require an extreme curvature of the zero isocline. Such an effect could be obtained only from a virtual threshold – a few individuals having the same practical effect as the total. The mechanism of competition of that kind is not easy to visualize, and the evidence is strongly against detectable interspecific competition between the two species that compete with each other and the remaining five species of Plethodontidae.

These examples of tests of the assumption of universality of competition, plus another to be described in detail in Chapter 6, show that ecological differences among coexisting species are insufficient evidence on which to base any broad conclusions. It should be noted that in all five of these studies, there was a priori reason to expect to find competition, and, if anything, there was bias in favor of a positive result. This effect has been discussed by Connell (1983). We need many more experiments, specifically with the designs published in advance, before we can make generalizations about the extent and the nature of interspecific interactions.

The community matrix

As an extension of the ideas about limiting similarity formulated by MacArthur and Levins (1967), Levins (1968) proposed that the competition coefficients among a number of species could be used as the elements of a matrix. This community matrix could be used to make several interesting predictions about the composition of communities, such as the number of species that a particular set of competition coefficients would permit to coexist, and the importance of relative niche breadths.

One study was specifically designed to test an implied prediction of the community matrix. This was Vandermeer's experiment on ciliate protozoans (Vandermeer 1969). He grew each pair of four species together in the laboratory and used the dynamic form of the Lotka–Volterra equations to calculate the competition coefficients. These were in turn used to predict the course of population histories of the four species when they were grown together. Specifically, Vandermeer assumed correctness of the model as far as the two-species systems were concerned. The question to be answered from the four species together was whether or not there were interactions among the species that were not accounted for by the observed competition coefficients. He called these "higher order interactions." They can be thought of as representing either coalitions of two species increasing the effect on a third species or some form of beneficial effect between two species reducing the negative effect of a third. Vandermeer's results were that the two-species competition coefficients, used in the community matrix, gave predictions that were satisfying in correctly representing the trends of the separate populations, but that the numerical details were not accurately predicted. Specifically, two of the species were expected to become extinct in the four-species system, but one of them persisted. This experiment has been widely cited and has brought out at least one problem in using the community matrix. This is that it is possible to obtain false predictions from its literal use (Brenchley 1979). It has been pointed out that this result came from an unstable equilibrium (Thomas & Pomerantz 1981; Vandermeer 1981b). The problem posed by Vandermeer originally, of how accurately one should expect the model to predict community composition, remains, as does the question of the reason for its quantitative failure (Brenchley 1981).

Incidental to other experiments, higher-order competitive interactions have been detected by Hairston and associates (1968) and by Wilbur (1972), whose experiments on salamander larvae will be described in detail in Chapter 4.

Several studies have made use of the community matrix in such widely divergent communities as the original *Drosophila* species of Levins, aquatic cave arthropods (Culver 1970), marine invertebrates (Kohn 1971), lizards (Pianka 1969), birds (Pulliam & Enders 1971), and granivorous rodents (Brown & Lieberman 1973; Hallett 1982). Not all of these investigators specifically used the community matrix, but all used the MacArthur–Levins, the Hallett–Pimm, or the Schoener formula for the competition coefficient, and drew conclusions about coexistence from them. The most detailed studies were those of Schoener (1975), Roughgarden, Porter, and Heckel (1981), and Roughgarden, Heckel, and Fuentes (1983) on *Anolis* lizards in the West Indies. By comparing habitat use on islands with one, two, or four species and using the regression method, Schoener measured the intensity of competition between size classes and sexes of the species in the Bahamas and Jamaica. Roughgarden and associates compared habitat use, food, and microclimate for two sets of islands in the Lesser Antilles. They found evidence of niche partitioning to be strong in the northern group of islands, especially well documented on St. Maarten and St. Eustatius, but the resources were not partitioned in the same way on the two islands, even though each supported representatives of the same two species or species groups: *A. bimaculatus* and *A. wattsi*.

The same two islands were the locations of the experiments of Pacala and Roughgarden (1982, 1985). The experiments were designed to test for competitive effects of *A. wattsi* on the representative of the *A. bimaculatus* group (*A. gingivinus* on St. Maarten). The original papers should be consulted for details, but the data presented there and in Roughgarden and associates (1983) show that the numbers of each species introduced into experimental plots were between five and nine times the natural density. Thus, all plots were overcrowded intraspecifically, and the mixed-species plots were overcrowded for both species. Moreover, *A. wattsi* was introduced at densities nearly 2.5 times those removed on St. Maarten, and *A. bimaculatus* was introduced at densities twice those removed on St. Eustatius. The reciprocal manipulations were much less exaggerated. These manipulations biased the results toward an interspecific effect on St. Maarten and toward an intraspecific effect on St. Eustatius – exactly the direction of the original hypothesis. These facts make the results difficult to interpret, as far as community interactions are concerned.

The experiment was followed for about eighty days, with the following being monitored: distribution within the plots, height of individuals above ground, growth, egg volume, prey volume per lizard, and mean prey size. For all except the first, a significant effect of *A. wattsi* on *A. gingivinus* was claimed, whereas no such effect was found by *A. wattsi* on *A. bimaculatus*. The difficulty in accepting the results at face value comes from the severe intraspecific crowding of *A. bimaculatus*. After the removal of an average of 31.5 specimens, 60 were added to each plot. The lanes, which accounted for 55.6 percent of the fenced areas, were not acceptable habitat for this species. Thus, the appropriate number to replace would have been 14 – very close to the 13.6 per 64 m^2 reported by Roughgarden and associates (1983). The fourfold intraspecific density could account for the failure to detect any additional effect of adding *A. wattsi*, and the explanation that the lack of such an effect is related to resorce partitioning is unnecessary.

A more convincing experiment was conducted on a small islet near St. Maarten (Roughgarden et al. 1983). At low elevations on St. Maarten and on this islet, *A. wattsi* was absent, and correlative data suggested that its absence was due to competition from *A. gingivinus*. Two plots were established, and *A. wattsi* was introduced into both, after removing half of the resident *A. gingivinus* from one of them. During the first six months of the experiment, the survival of *A. wattsi* on the *gingivinus*-removal plot was twice that on the control. Hatchlings were also observed on the islet, giving strong evidence that the habitat was satisfactory in the absence of *A. gingivinus*. Competition appears to have been demonstrated in a convincing manner, although one could wish for replication.

I have described these experiments in detail in order to show that experimentation per se is not a guarantee that the question will be answered, and to point out some of the pitfalls involved in field experiments. Some of these experiments have also been cited as successful in prominent reviews (Schoener 1982, 1983; Connell 1983).

Refinements of the models

Stochastic variation

Despite claimed successes for the predictions of the models, specifically for the limiting similarity (May 1973a, pp. 165–71), it was recognized that a very restrictive assumption in the original form of the MacArthur–Levins approach was the complete determination of abundances by the matrix of competition coefficients. No allowance had been made for the effect of environmental fluctuations. May and MacArthur (1972) and May (1973b) investigated the theoretical effects of adding specified amounts of variance to the system. May concluded that as long as the variance was less than the value of the "largest real part of all the eigenvalues of the interaction matrix," the system would fluctuate within limits set by the ratio of the two terms (May 1973b, pp. 627–8), and thus could be considered stable, at least in a general sense. This relaxation of a restrictive assumption was welcome information to ecologists interested in communities, and further elaboration on this theme followed (Abrams 1975; Feldman & Roughgarden 1975; Goh 1976; McMurtrie 1976). There were important implications for the amount of overlap in resource use and for the relative sizes of the carrying capacities of competing species that permitted coexistence or permitted the addition of a species to an existing two-species system (May 1981b).

In the foregoing descriptions of models, various assumptions have been stressed. These have all been in the realm of the biological properties of the species concerned. They are summarized well in Heck (1976). There are also assumptions buried in the mathematics, and these have caused problems, even for highly sophisticated mathematical ecologists. A prominent example is provided by an analysis of May's work on the effect of stochastic variation (Turelli 1978). Turelli pointed out assumptions implicit in May's mathematical treatment. He concluded that May's stability criterion does not ensure stability unless restrictive conditions on the form of the model are satisfied. This analysis calls attention to a severe problem in community ecology. With few exceptions, mathematical ecologists are unsophisticated in the complexities and pitfalls of "real-world" observational or experimental ecology, and they frequently take a cavalier attitude toward the necessity of validating the assumptions in their models. Their inability to evaluate the quality of studies that they cite in such validations constitutes a serious impediment to progress. On the other side, biologically oriented ecologists are unable to uncover the assumptions contained within the mathematics of the models, and there is a strong tendency to accept the conclusions uncritically. This is especially true in cases where field data are consistent with the conclusions, but are also interpretable by other hypotheses.

Excellent examples of such problems are given by Dayton (1973). At superficial levels, two sets of his observations confirmed predictions of the competition-based niche-partitioning model of natural communities. In the first, three species of predators in the intertidal community of Washington State were found to have diets that overlapped in such a way that they formed a linear series. The "middle" species, the anemone *Anthopleura xanthogrammica*, had a diet that overlapped that of the starfish *Pisaster ochraceus* by an average value of 0.396, following the method of MacArthur and Levins (1967), and that overlapped that of another starfish, *Pycnopodia helianthoides*, by an average of 0.054. The two starfish species had diets that overlapped very little – average value 0.006. These observations conform to the theoretical predictions of MacArthur and Levins, because the overlap of the anemone was below the critical value of 0.544, above which diets should have converged, rather than remaining different. Further observations of the diet of *Anthopleura* in areas where *Pisaster* was experimentally removed and where the *Pisaster* density was artifically doubled showed that the similarity in diets was much greater than for controls in the area where *Pisaster* was abnormally abundant, and was less in the area from which it had been removed – just the opposite of what would be predicted from a theory based on competition. Close observations showed that *Pisaster*, in the course of feeding, dislodges some extra prey items, and these are washed into the tide pools and channels, where they fall prey to *Anthopleura*, which remains attached to the rock. Similar results hold for areas where *Pycnopodia* is common. Both starfish disturb the prey items that they have in common with the anemone, and the diet of the latter is facilitated by their activity, rather than restricted between them by competition.

In a second example, Dayton showed that although competition for space could be demonstrated among sessile organisms, predictions of the rise and decline of the number of species on cleared areas, based on competition, were verified for the wrong reason, and in fact took place because of invasion of the areas by predators after the initial colonization.

Renewable resources

The original Lotka–Volterra equations describing competition between species contained fixed limits to the number of individuals attainable. In view of the consumable and renewable nature of many limiting resources, such a restriction seemed excessive. MacArthur (1968) modified the equations by specifically incorporating the population dynamics of the resource and the use of that resource by the competitors. Schoener's use (1974b) of this approach in proposing a method of calculating the competition coefficients was described earlier.

A stimulus to this approach was provided by two papers on laboratory experiments on species of *Drosophila* (Ayala, Gilpin, & Ehrenfeld 1973; Gilpin & Ayala 1973). Their results showed that the original equations were inadequate to account for the data, because the final equilibrium population sizes required that the zero isoclines be curved instead of straight. Schoener (1976, 1978) produced a series of models, using terms directly relatable to the competitive process, that produced curved isoclines. In these models, the separate effects of exclusive and shared consumable resources are present, as are constants representing both intraspecific and interspecific interference. The most complex of these models contains eleven terms:

$$\frac{dN_1}{dt} = \left[\frac{I_{O1}}{(Y_1 + N_1 + \beta N_2)}\right]\left[\frac{I_{E1}}{(y_1 + N_1)}\right] - \gamma_{11}N_1 - \gamma_{12}N_2 - C_1 \qquad (1.7)$$

N_1 is the number of individuals of species 1; N_2 is the number of individuals of species 2; R_1 converts available energy into individuals of species 1; I_{E1} is input of energy exclusively used by species 1; I_{O1} is input of shared energy obtained by species 1; Y_1 and y_1 set limits to the consumption of energy by individuals; β is a conversion factor equating the utilization of I_{O1} by an individual of species 2 into the utilization by species 1; γ_{11} is the intraspecific density-dependent cost per individual; γ_{12} is the interspecific density-dependent cost per individual; and C_1 is the density-independent cost. The model has advantages in that it is easy to eliminate terms that are inappropriate to the system under consideration. Thus, if there is no interference, γ_{11} and γ_{12} can be dropped, as can I_{E1} when there is no energy source exclusively used by species 1. Schoener goes over the various possibilities and provides a table giving the mathematical consequences of various sets of conditions (models) (Table 1.3). Only if resources overlap completely, and there is no interference, are the zero isoclines straight. They curve for all other models, and they also intersect, although the intersections reflect unstable equilibria ("saddles") under some conditions. Despite the increased flexibility provided in this approach to interspecific competition, two important assumptions in the Lotka–Volterra derivatives are retained: All individuals have the same impact on the system, and the impact is instantaneous. The models pose stringent challenges to anyone attempting to use them on real situations, and I am unaware of any attempt to do so.

Table 1.3. *Comparison of intersections of zero isoclines under Schoener's models making the designated assumptions; positive intersections are those in the graphic quadrant in which all numbers are positive*

Model	Max. positive intersections	Min. positive intersections	Kinds[a] of intersections
Overlapping resources	0	0	
Overlapping and exclusive resources	1	0	n
Exclusive resources, inter- and intraspecific interference	3	0	n
Overlapping resources, intraspecific interference	1	0	n
Overlapping resources, interspecific interference	1	0	s
Overlapping resources, inter- and intraspecific interference	2	0	n, s, n − s

[a]s = saddle, n = stable node.
Source: Schoener (1978).

Models involving more than one trophic level

Most of the work on communities, both theoretical and practical, has involved competition between species at the same trophic level, although at least one of the models for estimating the competition coefficient [equation (1.5)] requires knowledge of the population dynamics of the "resource species." Despite growing evidence that predation is an important interaction in many natural communities (Huffaker 1958; Hairston, Smith, & Slobodkin 1960; DeBach 1964; Paine 1966, 1974), the incorporation of predator–prey interactions into community theory is relatively recent. The original Lotka–Volterra models were unsatisfactory in that there was no resilience – any perturbation was continued indefinitely. The conclusion of Leslie and Gower (1960) that density dependence in the prey population is a condition that gives damped oscillations was an important step in understanding predation. It was demonstrated experimentally by Luckinbill (1973) for *Paramecium* and its protozoan predator *Didinium* – the system that had previously failed to show persistence (Gause 1934; Hairston et al. 1968).

An early step toward incorporating predator–prey relationships into community ecology was the work of Roughgarden and Feldman (1975). They accepted the MacArthur and Levins (1967) array of three competing species ordered along a resource axis. To this model, Roughgarden and Feldman added a predator that exploits all three competing species. They were able to show that predation pressure could reduce the minimum niche-separation distances of the prey species (increase the amount of overlap of resource use) and, if strong enough, could permit complete resource overlap. This conclusion seems intuitively straightforward.

The extension of predator–prey relationships to community considerations has taken two routes: food-web theory and loop analysis. These have been pursued independent of each other.

Food webs

Diagrammatic food webs were used by MacArthur (1955) to illustrate his argument that diversity in natural communities enhances stability, but they were largely ignored in the subsequent two decades, while competition held center stage in theoretical community ecology. Cohen (1978) attempted an inductive approach to understanding food webs by looking for evidence of common patterns. He drew several conclusions from his analysis of a number of food webs that had been reported in the literature. The first was that there was a strong tendency for them to be representable by interval graphs. The importance of this finding was that it meant that niche overlaps among the species in the food web could be represented along a single dimension. Cohen constructed randomly assembled food webs and found that the real ones that he analyzed could be represented by interval graphs far more frequently than chance allowed. He pointed out that this was a sufficient condition, but not a necessary condition, for exploitative competition. He found that the number of kinds of predators (in his terminology, any species that eats another – herbivores would be included) was greater than the number of kinds of prey in the ratio 4 : 3 and that this ratio held closely throughout his community food webs. He distinguished between community webs, which considered all feeding relations within a habitat, and "sink" webs, which considered only the prey items of a selected group of predators, or "source" webs, which considered only the predators of a selected group of prey. In sink webs, of course, the ratio was much lower, because only some kinds of predators were considered. A third conclusion, that "every predator in a given food web has a constant and independent probability of preying on every prey in that food web" (Cohen 1978, p. 61), was controversial (Pimm 1982). Cohen pointed out a problem that is ignored by some workers. It is that the webs are most commonly reported as lines connecting predators and their prey, with no indication of the relative magnitudes of the interactions, or at best crude estimates. Analyses based on "plus–minus" relationships probably are not worth much, but quantitative data, in terms of the proportion of the predator's food represented by each kind of prey, or especially the proportion of the prey produced that are consumed by the predator, are not easily obtained. Nevertheless, firm conclusions about communities consisting of more than one trophic level will depend on such information.

Pimm (1982) took a different approach from that of Cohen. He began each question with a consideration of the possible interactions and tested mathematically for stability of the systems, arguing that possibilities that were not stable would not be observed, and hence did not merit further consideration. He then looked to nature for confirmation of his conclusions. Pimm listed twelve "causes" of the patterns in food webs, along with the methods by which the patterns are explained: dynamics (i.e., mathematically), "biology" (direct observations), or energetic considerations, as from successive multiplications of low ecological efficiencies. His conclusions are as follows:

1. There are no loops (*A* feeds on *B* feeds on *C* feeds on *A*), no predators without prey, and no singular systems (two predators exclusively on one prey).
2. There are limits to the complexity of food webs.
3. The lengths of food webs are limited.
4. Omnivores are uncommon.
5. Omnivores feed on adjacent trophic levels.
6. The two preceding patterns are not followed by insects and their parasitoids.
7. Scavengers do not distinguish among trophic levels.

8&9. Compartments (distinct parts of webs) only occur when in separate habitats.

10. There are more kinds of predators than of prey.
11. There is a negative relationship between the number of species of prey and the number of species of predators on a species.
12. Prey overlaps can be represented by interval graphs.

He regards items 7, 10, and 11 as uncertain. Most of the patterns were explained by "dynamics," few by energy. The paucity of omnivores must reflect the absence from his data sets of food webs containing birds of the Families Fringillidae and Icteridae, among others. Many of the species of these families are abundant and prominent members of their communities.

Food webs involve one glaring omission. There is no way to include interspecific interactions between the species at the lowest trophic level. Hence, competition among producers, a prevalent characteristic of terrestrial communities, cannot be included in the analyses. Moreover, exploitative competition is the only mechanism of competition allowable at any trophic level – interference cannot be included in the data. In his discussion of scavengers, Pimm omits discussion of the organisms that actually accomplish decomposition: bacteria and fungi. This surely distorts some of his conclusions about detritus-based food webs, inasmuch as he omits a trophic level. It is widely recognized that nearly all detritus feeders derive nutrients from the bacteria and fungi. Food webs based on carrion may be as he argues, but no examples are given, despite the fact that they are the most obvious cases of donor control. Moreover, the dynamics of such communities do not appear to depend on food-web characteristics (Beaver 1977; Kneidel 1984a,b, 1985).

Loop analysis

In an extension of the mathematical principles used in the community matrix, Levins (1975) provided the conditions for stability of any group of interacting species. The necessary information is the sign of the effect of each species on the other(s) and on itself. The species are placed mathematically in a matrix, but are visualized graphically as boxes connected by lines of positive or negative sign. Inasmuch as a species may be self-limiting or positively affected (as in the Allee effect), a line may connect back to its starting box. Thus, a loop of effects may involve 1, 2, 3, . . . , n boxes taken in order. The sign of the loop as a whole is the product of the signs of the connecting lines. These are the feedback effects of the interspecific interactions. All are negative in stable systems, but Levins points out that for systems of realistic complexity, the magnitude of the effect may need to be known.

Hutchinson (1978) gives the example of a three-level "community," consisting of a predator, a herbivore, and a self-limiting plant. As a loop diagram, it would be represented as in Figure 1.1. From Levins's description of the method, the requirements for stability are as follows:

1. That the feedback of each level of complexity be negative, as follows: (a) F_1 Σa_{ii} must be negative. (b) At the level of two connected species, F_2 $\Sigma a_{ij}a_{ji}$ $-$ $\Sigma a_{ii}a_{ji}$ must be negative. (c) At the level of three species, F_3 $\Sigma a_{ii}a_{ji}a_{kk} - \Sigma a_{ii}a_{jk}a_{kj} + \Sigma a_{ij}a_{jk}a_{ki}$ must be negative.

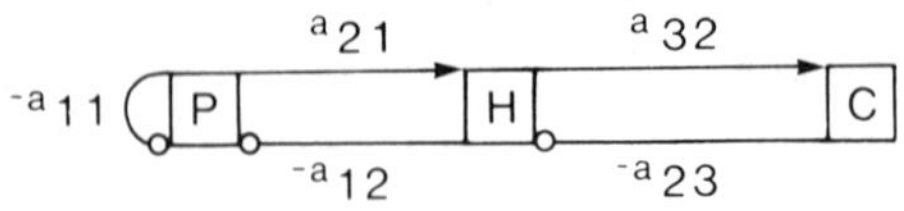

Figure 1.1. Loop diagram of a simple community with a self-limited plant species (P), a herbivore (H), and a carnivore (C). Arrows show positive effects; lines ending in circles show negative effects. The symbols follow the convention that the subscript shows the effect on the first species by the second. Thus, a_{21} means that species 1 has a positive effect on species 2.

2. The feedback from the long loops cannot be greater than the product of the feedback effects at the two lower levels: $F_1F_2 + F_3 > 0$.

In the present example, all of the conditions for stability are met:

$$F_1 = -a_{11}$$
$$F_2 = -a_{12}a_{21} - a_{32}a_{23}$$
$$F_3 = -a_{11}a_{23}a_{32}$$
$$F_1F_2 + F_3 = a_{11}a_{12}a_{21} > 0$$

Now, one of the most attractive features of the loop analysis is the need to know only the signs of the effects, but it turns out that in some of the examples given by Levins, the magnitude must also be known. From the practical standpoint, the detection of an effect may be difficult, as seen in the examples of field experiments on lizards, salamanders, and beetles cited earlier. Measuring the magnitude of an effect is much more difficult, and furthermore frequently requires assumptions that may or may not be valid. A good example is my salamander experiment (Hairston 1980b, 1981) in which the calculation of competition coefficients assumed the linearity of zero isoclines, as did the claim that some species were unaffected by the experimental removal. The exchange between Schoener (1983) and me (Hairston 1985) shows not only the importance of the model used but also the quantitative properties that must be known when using some models. This was a carefully done, long-term experiment. If the results cannot be used with confidence in a loop analysis, I argue that loop analysis, while stimulating intellectually, is of practical use only in very simple systems, where the interactions have already been determined experimentally.

Null models

Null models have achieved prominence in recent literature, but the principle is not new. I first consider an early null model and then describe some newer examples and the controversy surrounding them. The first community model to be challenged by a quantitative null hypothesis was a simple verbal one conceived by Charles Elton. It is of some interest that the first null models were proposed by C. B. Williams (1947, 1951), who had spent much effort on correlations between the abundance of insects and the influence of various components of weather. He first questioned the conclusion of Elton (1946) on the importance of his observation that the number of species per genus within communities was consistently less than the number in the large area of which the communities were parts. Elton argued that the difference was due to competition between closely related species inhabiting the same community. Williams pointed out that the relationship between the number of species per genus and the inclusiveness of samples of a complete fauna was not a linear one. Using a model (the logarithmic series) that he had devised earlier with the help of R. A. Fisher (Fisher et al. 1943), he claimed that the expected proportion of congeneric species within the communities used by Elton was actually less than was observed. Williams concluded that the tendency for similar species to have similar requirements and therefore to be found together was stronger than any effect of interspecific competition.

Williams was aware that his conclusions depended on the assumption that the logarithmic series was the proper model on which to base expectations. In a later paper (1951), he used a different model in performing the same kind of analysis on a study of East African birds by Moreau (1948). Moreau had objected to the large sizes of the communities used by Elton and by Williams, and his study involved subdivisions of major habitats. He concluded that ecological segregation among related species was an important feature of his study, and "that most of the cases of habitat overlap apparently do not involve overlap in diet and hence are not cases of complete ecological overlap and do not involve competition" (p. 117). In analyzing Moreau's data, Williams avoided the logarithmic series and resorted to a model that involved no assumptions about frequency distributions. The number of pairs of species in any assemblage is simply $N(N - 1)/2$. The number of congeneric pairs depends on the number of genera with 1, 2, 3, . . . , n species; it is $\Sigma n_i(n_i - 1)/2$, where n_i is the number of species in each genus i. Thus, the whole area provides the expected case for the situation within any habitat, because the chance that two species selected at random are congeneric is $\Sigma n_i(n_i - 1)/N(N - 1)$. In a clear statement of the purpose of a null model, Williams says: "I think that it will be of value to study his data from a statistical point of view, because no attempt was made to show whether or not the number of such cases was larger or smaller than might be expected by chance; and also because Moreau's data, being the results of observations by a field ecologist who is familiar with his terrain and his birds, are likely to be as good as it is possible to get anywhere for such an analysis" (Williams 1951, p. 247). His calculations showed that the number of congeneric pairs of species expected by chance within habitats was 16.5, whereas the number observed was 48. The number of pairs of confamilial species observed within habitats was 277, whereas the expected number was 91.5. For both congeneric pairs and confamilial pairs of species, there were three times as many within habitats as would have been expected by random selection from the whole fauna. Analyses of combinations of three or more species showed more extreme results, and Williams repeated his conclusion that the tendency of similar species to have similar requirements was stronger than the influence of interspecific competition.

The rigor implied in that statistical analysis was refreshing in a field plagued by post facto reasoning, and the avoidance of assumptions about mathematical frequency distributions made the approach especially appealing. I therefore applied the analysis to my own data on plethodontid salamanders in the southern Appalachian Mountains (Hairston 1949, 1951). I had obtained evidence that I interpreted as indicating that interspecific competition had been important in structuring the salamander communities of that region, having found that congeneric species either tended not to be found together or else were differently distributed over several habitats. To my surprise, the analysis gave the same results as Williams had obtained for the birds.

Understanding came when I realized that although no mathematical assumptions were involved, there were three biological assumptions that had gone unnoticed (Hairston 1964):

1. The number of different habitats occupied by a species is unrelated to the number of species in that genus.
2. The presence of a species in a habitat can be interpreted without reference to its presence or abundance in other kinds of habitats.
3. The human observer divides the total area into habitats in the same manner as do the species being studied.

All of these assumptions lack validity.

The implications of the first assumption can best be understood by considering the monotypic genera in any larger group of organisms. The reason for the recognition of a monotypic genus is its distinctness, a feature associated with specialization. The expectation would be for such a species to be confined to a single kind of habitat. If the species of larger genera are less specialized, they would be expected to occur in more than one habitat, and that would force upon the data a bias in the direction found by Williams. The variability of species of large genera was noticed by Darwin, who devoted part of Chapter II of *The Origin of Species* to the phenomenon. He was thinking about morphological variability, but it is also true that variability in ecological tolerances is more widespread among species in large genera than it is among monotypic genera. Among the salamanders of the Subfamily Desmognathinae, two genera are monotypic, and one genus has eleven species. The monotypic genera are confined in habitat, as well as in geographic distribution. *Leurognathus* is completely aquatic and occurs only in swift mountain streams in the southern Appalachians; *Phaeognathus* is fossorial and is restricted to a part of the Red Hills formation in south central Alabama, where it occurs in shaded, moist ravines. In contrast, most species of *Desmognathus* can be found in two or more habitats. *Desmognathus monticola*, *D. fuscus*, *D. auriculatus*, and *D. santeetlah* can be found in the water of streams and seepage areas, on saturated, but not inundated, parts of streambeds, and to some extent in the immediately adjacent part of the forest (up to 6 or 7 m from water). They thus qualify as being in three habitats. *D. ochrophaeus* and *D. imitator* are found in the same three habitats and range much farther into forests, which include both spruce-fir and deciduous types. *D. quadramaculatus* is more aquatic than those already described, but some large individuals can be found well away from water, especially at night. *D. brimleyorum* and *D. welteri* are apparently equally aquatic. Two species are terrestrial, with no larval stage. *Desmognathus wrighti* is a high-altitude species. It is typical of spruce-fir forests, but is also a regular inhabitant of hardwoods at elevations above 1,000 m. *D. aeneus* is confined to the vicinity of streams in hardwood forests, but has an altitudinal range from near sea level to 1,370 m.

The impact of these statistics can be seen by considering the community shown in Table 1.4. There are four habitats that are recognizable independent of the salamander distributions: the streams, the streambanks, the deciduous forest, and the spruce-fir forest. *Leurognathus*, the monotypic genus, is found only in the streams. Its inclusion on the basis of only four specimens is justified by the fact that all other specimens ever collected were found in the same kind of habitat. Among the four species of *Desmognathus*, *D. quadramaculatus* and *D. ochrophaeus* are found in all four habitats; *D. monticola* is absent from the spruce-fir; and *D. wrighti* is found only in the two forest habitats. Thus, the area contains $(5 \times 4)/2$, or ten pairs of species, of which $(4 \times 3)/2$ are congeneric – 0.6 of the total. In the water, four species are found (6 pairs, 3 congeneric); on the streambanks, there are three species (3 pairs, all congeneric). The same is true of the spruce-fir forest. In the deciduous forest, there are four species, all congeneric. There are thus 6 + 3 + 3 +

Table 1.4. *Ecological preferences in the* Leurognathus–Desmognathus *series (percentages)*

			Distance from stream (ft)						
Species	Stream	Streambed	1–10	10–20	20–40	40–60	60–100	100–200	>200
L. marmorata (4)[a]	100	0	0	0	0	0	0	0	0
D. quadramaculatus (159)	78.6	19.6	1.2	0.6	0	0	0	0	0
D. monticola (138)	21.8	54.4	19.0	4.1	0.7	0	0	0	0
D. ochrophaeus (379)	6.9	16.3	22.5	18.0	8.7	4.2	2.4	4.2	16.8
D. wrighti (50)	0	0	0	0	4.0	6.0	6.0	8.0	76.0

[a]Total number of specimens on which percentages are based.
Source: Hairston (1949).

6 = 18 pairs, of which 3 + 3 + 3 + 6 = 15 are congeneric, or 0.833 of all pairs in the separate habitats. This is 39 percent more than random selection from the area as a whole would lead us to expect. If the monotypic *Leurognathus* were not more restricted in habitat than the *Desmognathus* species, it would be found in three habitats, and the habitats would, as a group, be like the total area. It is the difference between the species of the monotypic genus and the others that brings about the statistical agreement with Williams's analysis of Moreau's data.

Thus, although some species of the fairly large genus *Desmognathus* are confined to a single habitat each, most of them are not, the mean number of habitats being 2.6 per species. One could review the remaining five monotypic genera and the remaining five larger genera of eastern Plethodontidae with similar results.

Desmognathus also provides a good illustration of the nonvalidity of the second biological assumption contained in Williams's model. As many as five species of this genus occur together in adjacent habitats in the southern Appalachians. In a four-species array where the ecological distributions were recorded, they were *D. quadramaculatus*, *D. monticola*, *D. ochrophaeus*, and *D. wrighti*. Each habitat has three or four species, but each species had a characteristic frequency of occurrence in each habitat (Table 1.4). The differential distributions imply that the species have ecological impacts on each

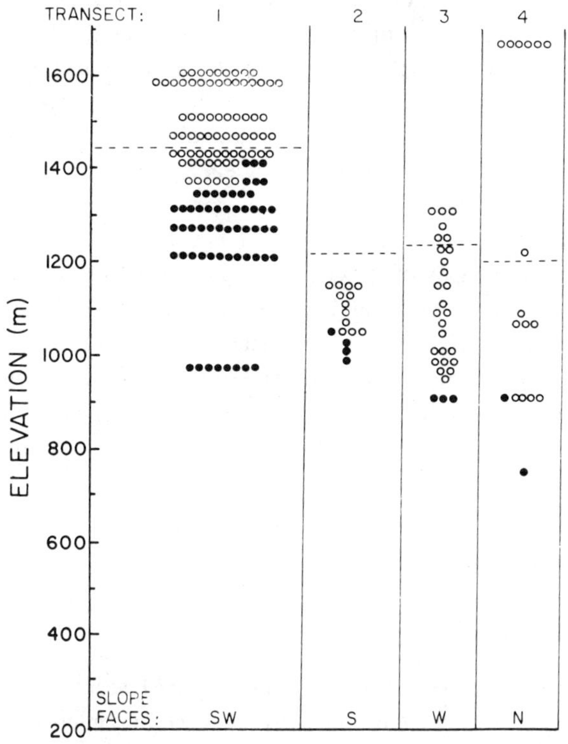

Figure 1.2. Altitudinal distribution of *Plethodon jordani* (open circles) and *P. glutinosus* (filled circles) in four transects in the Great Smoky Mountains. Broken lines show approximate lower limit of spruce-fir forest.

other (see Chapter 6 for a treatment of these interactions), but Williams's analysis would show 3, 3, 6, and 3 congeneric pairs in the four habitats, because the different abundances of the species in the different habitats would be ignored, as would their distributions within the forest.

The third assumption implicit in Williams's null model is shown to be invalid by the altitudinal distributions of two species of the salamander genus *Plethodon*. Figure 1.2 shows that *P. jordani* and *P. glutinosus* have nearly exclusive distributions in the Great Smoky Mountains. The altitudinal overlap is no more than 120 m on any of a large number of transects that have been made. The original interpretation, now fully confirmed by removal experiments (see Chapter 5) was that the two species are in intense competition, each one being superior at its appropriate elevation. In all cases, the overlap zone lies within the hardwood forest, and if we accept the observer's classification of habitats, both species must be included in that habitat, thus implying minor importance to competition according to the null model.

I have discussed this example of a null model in detail because it is especially clear and free from mathematical assumptions and because it illustrates a few of the difficulties encountered when null models are constructed. It is not always so easy to identify the biological and mathematical assumptions in any model, and null models are not exempt from the problem. Some of the implications of this analysis are clear. First, it is not proper to use the generic diversity, or the lack of it, within and among habitats to make interpretations about the importance of interspecific competition in structuring natural communities. Notice that this implication is broader than either the original model or the null model. Second, we should be cautious about the use of diversity, however defined or measured, to compare communities, especially if the comparison is used to draw conclusions about the strength of interspecific interactions.

Derived models

Size relationships

In the original paper that inspired much of the research on ecological communities and guilds, Hutchinson (1959) proposed that interspecific competition would prevent the coexistence of related species that were very similar in terms of the sizes of structures that reflected their feeding habits – bird beaks or mammal jaws, for example. This hypothesis became extremely popular, and many papers were published in the ensuing twenty years affirming that there was a minimum ratio of sizes below which potentially competing species did not coexist and that there was a constancy in the ratios of sizes of coexisting related species. The ratio was thought by most investigators to represent the limiting similarity of MacArthur and Levins in an evolutionary sense.

Coexisting species from many different taxonomic groups have been examined from this standpoint, with few authors reporting that size ratios failed to meet the expectations. An extensive list of citations is given by Simberloff and Boecklen (1981). It is tempting to accept these minimum size ratios, especially because they agree with theoretical developments since 1959. What is impossible to discover is the number of examples that do not agree with theory. Few ecologists would have attempted to publish such observations, and it is likely that fewer editors would have accepted such negative results for publication. What we see, then, are highly selected data – obtained by the process (largely unconscious, no doubt) of "data dredging." The implications of this practice have been discussed by Selvin and Stuart (1966), Simberloff and Boecklen (1981), and Connell (1983). It is not a straightforward matter to assess directly the scientific significance of a selected observation without knowledge of the probability that the observation was due to chance.

Two different approaches have been taken to make just such an assessment. Roth (1981) proposed two statistical tests of the constancy of size ratios and found support for the phenomenon in a New Guinea fruit pigeon guild (Diamond 1975, 1978). She cautioned, however, that this was a highly selected example. Of more interest is her analysis of the large number of size ratios among congeneric bird species reported by Schoener (1965). These by no means show constancy or even a central tendency, but are readily described by a declining exponential relationship. Roth asserted that it would be possible post hoc to construct an elaborate model mimicking the exponential decay curve, but it seems more likely that the following simple explanation will suffice: Closely related species are likely to be very similar in size, and the probability of a difference should decrease as the difference increases.

Almost simultaneously with the publication of Roth's paper, Simberloff and Boecklen (1981) published a null model to test the statistical validity of a large number of studies claiming either a minimum size ratio or a constant ratio among coexisting closely related species. Among the thirty-one studies examined, twenty-one claimed constant size ratios, eighteen claimed minimum size ratios, and three claimed neither. The arithmetic discrepancy comes from the fact that eleven studies made both claims. Among the twenty-one claims of constant size ratios, only four were fully sustained statistically, with an additional three receiving some support. The analysis of one study revealed that there were significantly too many small ratios – exactly the opposite of the claim. It is of special interest here that one of the examples supported statistically involved salamanders (Krzysik 1979). That this size relationship may not be closely related to competition is documented in Chapter 6. A single study among the eighteen claiming a minimum size ratio could be fully supported statistically, and two showed significantly small ratios. Thus, of thirty-nine claims, five were supported in a statistically acceptable manner, and three revealed the exact opposite of what was claimed. The only surprise in this outcome is the extent to which the bandwagon spirit pervaded the field. That size differences are not uniformly consistent with the original hypothesis is to be expected. As Schoener (1965) pointed out, there are too many different resources for which species can compete and through which natural selection can operate to bring about sufficient differences for coexistence. The idea of a minimum or a constant size ratio is far from being sufficiently general to be a "law of nature."

Biogeographic relationships

MacArthur and Wilson (1963, 1967) brought community ecology into biogeography with their equilibrium theory. This takes into consideration the effects of competition and other interspecific interactions in determining the species compositions of island biotas. The theory revived interest in biogeography and stimulated ecologists to apply community theory to biogeographic questions. The universal question is the extent to which interspecific competition has influenced the pattern of distributions on archipelagos.

The most thoroughly worked system is that of the geospizid finches of the Galapagos Islands (Lack 1947; Bowman 1961; Grant 1981). Two kinds of data have been used to draw conclusions about the effects of competition in determining the structures of the island communities: the generic compositions of the communities and the relative sizes of the beaks of coexisting species. The most interesting feature is the fact that thirteen of the fifteen islands have more of the fourteen genera than would be expected if the twenty-three species were distributed randomly over the archipelago (Harris 1973; Simberloff 1984). This is not the usual pattern for island groups, where congeneric species tend to occur more frequently on individual islands than would be expected by chance (Connor & Simberloff 1978). The pattern among the Geospizidae is consistent with the hypothesis that congeneric species tend to exclude each other competitively, but that species of different genera have little mutual effect (Grant & Abbott 1980).

A related observation concerns the small fraction of the possible combinations of species in the genus *Geospiza* that actually occur, with cooccurrences of three, four, or five species on one island being significantly less frequent than expected. The alternative hypothesis of differential physical suitability of the islands for the different species of *Geospiza* (Pielou & Pielou 1968) is possible, but does not explain the concomitant observation that beak sizes among coexisting species frequently show more variation than randomly assembled representatives of the species would lead us to expect (Simberloff 1984). Thus, in this example, the importance of interspecific competition in determining community structure seems well established.

It is less clear that the avifaunas of other archipelagos have been as strongly influenced by competition. Diamond (1975) produced an elaborate analysis of the birds of the Bismarck Archipelago adjacent to New Guinea. His interpretation of the data from more than fifty islands was that competition had been a powerful force in determining which combinations of species occurred on which islands. He called his series of statements "Assembly Rules."

Connor and Simberloff (1979) challenged his conclusions on the grounds that Diamond had not given sufficient attention to the alternative null hypothesis that the patterns that he was attempting to explain could have occurred by chance assembly from the large pool of 147 species of land birds. They could not obtain the original data, but did test their null model on data from the New Hebrides, east of the Bismarcks, and on the birds and the bats of the West Indies. For the birds of the New Hebrides, they found a close match between the observed numbers of pairs of species, whether confamilial or not, and those expected from a chance assembly from the pool of fifty-six species. For the other faunas, the match was not bad, but did show the observed and expected occurrences to be significantly different. They concluded that Diamond's assembly rules were unjustified. They did not deny that competition is important in bird communities, but they did argue that the evidence was inadequate and the analysis faulty.

Diamond and Gilpin (1982) responded to the arguments of Connor and Simberloff. In addition to statistical arguments, which are complex and not resolved at present, they pointed out that Diamond's analysis involved separating the species into guilds, and thus the Connor and Simberloff tests had "diluted" presumably interacting groups with species that were unrelated ecologically. This is a valid argument, assuming that the assignment of species into guilds was valid. It was based on similarity of feeding habits, a practice that will be questioned later. After the response, there followed a further exchange (Connor & Simberloff 1984; Gilpin & Diamond 1984). This literature makes unpleasant reading with its mutual recriminations. The unbiased reader (assuming that one exists) is left with the feeling that the bitterness of a scientific dispute is inversely related to the probability of a clear resolution.

One crucial point is the accuracy with which guilds can be delimited. Earlier in this chapter I described my experiments on a group of salamander species that appear to conform to the usually accepted definition of a guild. Removal of either of the two most abundant species had no detectable effect on the remaining five species, thus raising the serious question whether or not guilds can be delimited a priori without experimental evidence of the supposed competitive basis. A second group of salamander species, among which there was apparent evidence of niche partitioning, has been found to be structured largely by predation (Hairston 1986; see Chapter 6), and I wonder how many of the "guilds" of birds of the Bismarck Archipelago would be supported by an experimental test.

The problem of identifying communities and guilds

It is an interesting circumstance that a rich and coherent theory of community organization has been developed over the past two decades, without a specific statement of what the limits of a community are, or how communities can be distinguished from each other or even from other similar groups of species. It is true that definitions have been advanced throughout the history of ecology, as discussed later, but most of the definitions have been ignored in the development of community theory. It is also true that in nearly every treatment, the author's implicit assumptions about the nature of a community can be understood from the context. This does not mean that any a priori definition entered into the selection of the group of species under consideration.

The explanation of what we observe in nature is, of course, a legitimate part of ecological science. It is a step toward a complete science, but we should work toward the step of prediction, which is an integral part of any science. Prediction implies prior knowledge of the initial conditions, and if we are to predict events within a community, the initial composition of that community must be known in advance.

I now consider definitions of communities that have been offered by ecologists and the implications of these definitions for the quantitative models that have already been described. Although the distinction has not always been clear in the literature, the term "community" will be distinguished from the term "guild."

Community definitions: early meanings

Until midcentury, the term "community" embraced either all of the plants or all of the organisms in a delimited area. The pioneer ecologists Clements and Shelford (1939) stated that: "ecology is in large measure the science of community populations" (p. 3). Their concept was further elaborated: "As has been earlier emphasized, the cause-and-effect cycle in the biotic community comprises the action of the habitat upon the associated organisms, their response to this, and the consequent effect upon the physical factors of the habitat" (p. 68). The emphasis in community ecology during that period was on the intricate ways in which the multitude of species were interdependent. The attitude culminated in the definition of a "major community" by Allee and associates (1949, p. 436): "In large, the major community may be defined as a natural assemblage of organisms which, together with its habitat, has reached a survival level such that it is relatively independent of adjacent assemblages of equal rank; to this extent, given radiant energy, it is self-sustaining." Earlier, Tansley (1935) had combined the community and the nonliving part of the environment into the single term "ecosystem" – a term that is now universally accepted. The most notable attempt to quantify the relationships within one of these assemblages was that of Lindeman (1942). His emphasis was entirely on energy flow and hence on the food-web aspects of interspecific relationships, as shown by his title: "The Trophic-Dynamic Aspect of Ecology." Competitive relationships between species at the same trophic level were not central to thinking about communities.

An extreme example of a community definition, based on acceptance of the universality of interdependence of coexisting species, was that of Dice (1952, p. 15): "An ecologic community is an assemblage of ecologically related organisms composed of 2 or more species. Such a community may be of any ecologic rank and may include any number of associated individuals." His illustration of a community was a rabbit under a blackberry bush. No ecologist would deny that there are mutual interactions between the rabbit and the blackberry bush, even though direct feeding is probably not involved. Nevertheless, it would be difficult to make a quantitative assessment of the strength of any of the proposed interactions.

Restrictions on community inclusiveness

Implicit reductions in the generality of the array of species covered began with analyses of the relative abundance of coexisting species. Fisher and associates (1943) analyzed the relationships among macrolepidoptera caught at light traps, and Preston (1948) considered all coexisting birds in his analysis of the same general kind. These were ad hoc mathematical descriptions – quantitative rather than mechanistic advances. The approach of Fisher and associates had the primary purpose of measuring diversity in a way that permitted comparisons between communities. Preston's approach was similar, but also carried the argument that the "universe" is a coherent unit of a size fixed by the numerical relationships. He did not use the term "community." MacArthur's broken-stick models proposed that the kind of competition among coexisting species could be detected by the match of an observed species abundance distribution to one of several mathematical models based on biological premises (MacArthur 1957, 1960). Originally, these models were applied to all coexisting birds, but shortly the taxonomic level became even more restricted, in some cases to a single genus (e.g., Kohn 1959).

Thus, in two steps, the conception of the community changed from an all-embracing food web to a single trophic level, and thinking about the interspecific interactions became restricted to competition. What had been community ecology became ecosystem analysis, and groups of competing species came to be thought of as communities. Such was the usage when Levins (1968) devised the community matrix, which consisted of the mutual competitive interactions among a group of species. Until these developments, the only problem in delimiting communities had been that of how much physical space was required for the contained organisms to be considered a complete community. Naturally, assemblages situated on environmental gradients caused operational difficulties in separating them into communities, difficulties that were described by Curtis and McIntosh (1951) and by Whittaker (1952).

The new approach to communities did not necessarily avoid the problem of physical delimitation, and it did create a new one: how to decide on the inclusiveness of the competing group. The issue was clarified, but not settled, by coining a new term: "guild" (Root 1967).

The guild and its complications

Recognizing that "community" was a term whose meaning had become confused, Root (1967, p. 335) proposed that one of the meanings be taken over by the term "guild": "A *guild* is defined as a group of species that exploit the same class of environmental resources in a similar way. This term groups together species, without regard to taxonomic position, that overlap significantly in their niche requirements." Taxonomic restriction was thus specifically avoided, but the extent of the class of environmental resources was left unclear. The problem can be visualized by considering the fact that many birds and many parasitoid insects share the common resource of (largely) herbivorous insects in forest habitats, yet nearly all publications on guilds concern groups of species belonging to the same taxonomic class of organisms, or are more closely related. The guild concept has been popular for several reasons. First, as usually interpreted, it allows the investigator to assess community interactions among a limited number of species. Second, the kind of interaction is, almost by definition, restricted to interspecific competition. Third, as supported by generally accepted theory, it has been considered necessary only to provide evidence that is compatible with that theory, rather than using the opportunity to ask whether or not the theory is valid. The inclusiveness of guilds has been a problem throughout the history of the use of the term. The result has been that operationally, guilds have been delimited in a highly subjective manner. The consequences of this practice were discussed by Jasic (1981). The problem can be illustrated by two lists of guilds from the literature (Adams 1985; Taper & Case 1985). It is important to understand that in both studies the authors were using their selected guilds to support niche partitioning as the basis of guild structure. Adams (1985) lists thirty guilds; Taper and Case (1985) list thirty-one; but there is not a single example common to the two lists. There is a systematic difference in the inclusiveness of the guilds listed. Taper and Case include none with more than five species, and the average number of species is 3.03; Adams includes sixteen examples among his thirty guilds with more than five species, and the average number of species is 7.37. In the Taper and Case table, twenty-five (81%) of the guilds are composed of congeneric species; in Adams's table, only eight (21%) of the thirty-eight guilds are composed of congeneric species (the discrepancy in number of guilds is due to the fact that not all studies listed single guilds). This is not to suggest that I can point to any of the sixty-one guilds as inappropriately chosen, but to emphasize the necessarily subjective nature of the choice.

In addition to the problem of the common restriction in practice to a single small taxonomic group, the guild concept implies knowledge that the shared resource is limiting the abundance of the guild members and that they are therefore in competition for that resource. If that is not the case, the guild is of much less ecological interest. For at least some guilds there exists evidence that the populations are limited for reasons unrelated to the common resource. A clear example, backed by experimental evidence, is Strong's study (1982) of leaf-eating hispine beetles on *Heliconia*. These species conform well to the definition of a guild, but Strong found no evidence of competition between them and concluded that their populations were limited by parasitoids. Another example, that of the terrestrial salamanders of the forest floor, was given earlier, and the example of the genus *Desmognathus* will be described in detail in Chapter 6.

Synthesis

Community ecology developed a strong theoretical base over a fairly short period following the stimulus of G. E. Hutchinson at the end of the 1950s. Coming at a time when ecologists felt a need to progress from what had been a largely descriptive science, the formulations of MacArthur, Levins, and others became popular as bases for explaining patterns that were readily discernible in field data. Despite theoretical arguments and the results of some field experiments showing the importance of predation, the entire edifice of theory was built on the assumption, overt or implicit, that competition for resources is the all-important force determining the ways in which communities function and in which the structure in terms of species composition and abundance is produced. The popularity of the theory is based on the number of observations that can be shown to be consistent with implied predictions. These concern ecological differences between coexisting species, distributional patterns, and the outcomes of a large number of experiments. Field experiments in which competition could not be detected, or that revealed that predation was responsible for determining community structure, were generally ignored.

It is true that several refinements of the basic models were provided by the theoreticians. These involved incorporation of the potential effects of stochastic variation and inclusion of the dynamics of renewable resources. These refinements raised more problems than they solved. The stochastic model was shown to include mathematical assumptions that greatly restricted its applicability. Models separating differential utilization of resources by competing species from other kinds of negative interactions, while accounting for experiments that showed the inadequacy of the simple Lotka–Volterra model, revealed that their application requires amounts of information that have so far rendered them impractical.

The inclusion of more than one trophic level into community theory has been proposed from two approaches: loop analysis and a theory of food webs. Loop analysis is deceptively attractive, as it purports to require only the sign of an interaction for its application, but for realistic amounts of complexity, the quantitative level of the interactions must also be known. Some interesting claims have been advanced from consideration of food webs, but their establishment remains to be demonstrated. There is one debilitating failing of food webs as data bases. They do not incorporate data permitting an assessment of competition within the lowest trophic level. As this is an important feature of terrestrial ecosystems, and also of some aquatic systems, such as plant-dominated communities on marine hard bottoms, it appears that food-web analysis will be of limited value in understanding nature.

The results from field experiments are accumulating at an increasing rate, and eventually they will force reassessments of community theory, but their impact is gradual, and it will require great insight to incorporate them into any general theory. A much more immediate impact on the field came from challenges to some of the derived applications of community theory. The assertion has been advanced that the advocates of regularities among size ratios of potentially competing species have ignored the possibility that the claimed pattern is the result of chance and that similar considerations apply to the patterns of distribution of related species on archipelagos. In the case of size ratios, the assertion appears to be justified. An important majority of claimed examples of constant or minimum size ratios fail to meet statistical requirements. In the case of island distributions, one, that of the geospizid finches of the Galapagos, has been found to show a pattern consistent with interspecific competition and to be far from a random array. Other examples are still subjects of intense debate, notably the birds of the island groups east of New Guinea, and it does not appear that the debate will be easily resolved.

The literature contains few, if any, examples of statements of the criteria by which communities can be recognized *before* research is started on them. A community could be all of the organisms in an area, or all of the coexisting species within a genus, or all of the coexisting species within a family, or any other taxonomic group, depending on whose usage is being followed. Such usages have always implied ecological interactions between the species involved. In recent years, the interaction has been assumed by most authors to be competition, especially where theoretical work has been concerned. This has led to the use of the term "guild," which has always implied competition for a common and usually known resource. When we look for evidence for the existence of communities or guilds as defined by the foregoing implications, we find that in a distressingly large proportion of cases there is a strong element of circularity. The ecological and morphological differences between coexisting species were demonstrated long ago. They lead us, through precise mathematical theory, to expect more of the same. It should not surprise us to find these expectations met, followed by the claim that the theory has been confirmed. One of the themes of this book is that such correlative evidence is unsatisfactory and that field experimentation is far superior to evidence for niche partitioning as a method of testing theory.

Recent years have thus brought community ecology into turmoil, partly because of the nature of the organisms to which the theory has been applied. Some of them, birds in particular, though seeming to conform to the assumptions on which the theory is based, are impractical from the standpoint of making a priori tests of the theory because experimentation is impossible or at best impractical. Others, such as many insect or other vagile invertebrate species, provide no reason to suppose that they meet the basic assumption of population equilibrium. In the following chapters, I intend to establish the fact that many species of salamanders do meet the assumptions of community theory and have been shown to be unusually suitable for the experimental manipulations that are the hallmark of acceptable scientific procedure. The experiments that have been conducted will be described in detail, and their relevance to ecological problems explored.

2

Equilibrium and stability in natural populations

The requisites of many mathematical theories of ecological communities include the assumption that the species populations are in equilibrium. Relatively few treatments have included an assessment of the degree to which violation of that assumption can be tolerated, in terms of altering the predictions of the theory. There have now been enough long-term studies of salamander populations to provide assurance that many species show fluctuations in numbers that are minor, relative to those of most other groups of animals. The long-term population histories of some of those other groups are described later.

Stability of salamander populations and individual longevity

This chapter first describes studies in which populations of salamanders have been followed long enough to detect fluctuations large enough to cause difficulties in applying community theory.

As is seen in Chapter 3, analysis of the life history of a species frequently requires the assumption of a stationary population. Such a condition is most easily achieved when individuals are long-lived. In this chapter I describe studies that have led to estimates of the longevity of a number of species, both in the field and in captivity. The purpose of the chapter is to establish the credentials of salamanders as providing suitable material for ecological studies that can test the theories.

Population stability

Knowledge of the amount of temporal variation in population numbers can be obtained only by counts or samplings repeated over a considerable period. As can be inferred from the descriptions of life histories in Chapter 3, there have been several studies conducted over a sufficient period to give reasonable assurance that most irregularities in population density have been observed. These studies have been made on species having a variety of life histories. The necessary duration for a study depends on the length of a

Table 2.1. *Breeding population sizes, sex ratios, and juvenile production of* Ambystoma tigrinum *from a breeding site in South Carolina*

	Breeding adults			Juvenile production	
Year	Male	Female	Sex ratio (M : F)	Metamorphosing juveniles	Juveniles per female
1978–9	60	44	1.4 : 1	1,041	23.66
1979–80	65	26	2.5 : 1	2	0.08
1980–1	21	8	2.6 : 1	1	0.13
1981–2	81	42	1.9 : 1	410	9.76
Mean	57	30	1.9 : 1	363.5	8.41

Source: Semlitsch (1983a).

generation. Thus, some of the populations described later in this chapter could be assessed for fluctuations in a relatively short time, because the generations are brief. Salamanders have long generations, there being an immature period of three to six years in many species. For stochastic phenomena to be assessed, studies of population trends should cover at least the minimum duration of a generation. The immature stages of many species are most sensitive to disturbances, and if they respond to such events as drought or cold (or their reverse), the result could be an age-class dominance of the population. That would have implications for the assumption of equilibrium, because individuals of different ages and different sizes frequently have very different effects on the rest of the community. For example, Morin (1983a), using an experimental density of *Ambystoma tigrinum* larvae lower than that observed in the field, found virtually complete consumption of several species of frog tadpoles. Differences in breeding densities within a pond could make a great difference in the survival of the prey species. It happens that Semlitsch (1983a) recorded just such differences in the numbers of female *A. tigrinum* entering a pond in South Carolina (Table 2.1). The number of females varied by a factor of 5.5 over a four-year period, and the number of successful larvae varied by three orders of magnitude.

The studies of long duration were made on species with terrestrial, streamside, and pond life histories. The array of habitats thus represented is sufficient to carry assurance that salamanders in general have stable populations. These studies will now be described.

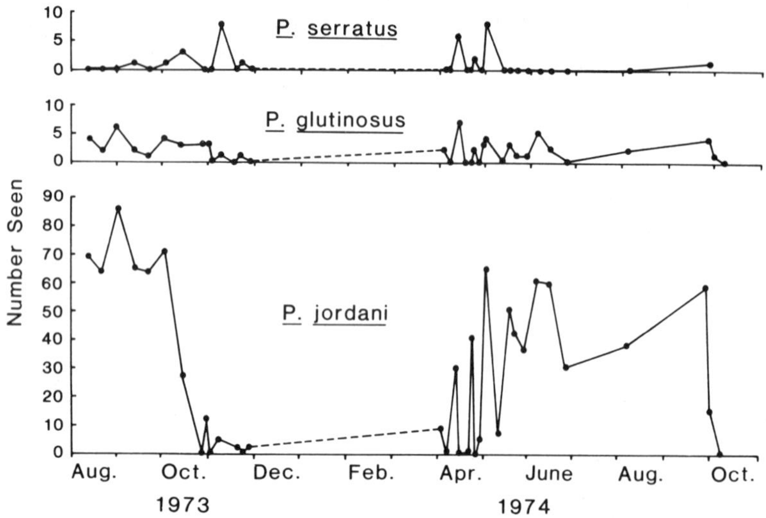

Figure 2.1. Seasonal changes in numbers of three species of *Plethodon* on an octagonal plot 24.384 m in diameter. Numbers are those individuals seen active above ground during one search of the plot.

Plethodon

Some terrestrial salamanders are active on the surface of the ground at night, sometimes climbing ferns, herbs, and low shrubs. Such species are accessible to direct counts of the numbers visible. *Plethodon jordani* and *P. glutinosus* are easily counted at night throughout the season of activity, which at the elevations where both occur is from the end of April until the first cold weather in autumn. Figure 2.1 shows the results of counts on the same marked plot over slightly more than one year. The plot was octagonal in shape, 24.384 m in diameter, and covered approximately 0.04 ha. The effect of a period of cold weather, when the air temperature fell below 0 °C, was dramatic. The effect lasted for the rest of the winter, as later in autumn there were warm rains that had no effect on the numbers of those two species seen above ground.

Seasonal activity patterns are not the same in all species of salamanders, even congeneric ones. The figure shows that *Plethodon serratus*, a species smaller than the other two, was less sensitive to the cold spell and virtually disappeared from the surface of the ground during summer. This seemingly anomalous situation was clarified by Jaeger (1980) and Bobka, Jaeger, and McNaught (1981). Working with *P. cinereus*, a close relative of *P. serratus*, they found that assimilation efficiency was inversely related to ambient temperature and concluded that *P. cinereus* had difficulty maintaining a positive energy budget at summer temperatures. The salamanders apparently avoid the problem by remaining below ground, where temperatures are lower than at the surface.

Although the data in Figure 2.1 and other counts of *Plethodon* to be reported later include all specimens seen, they do represent a sample of those present. That is because not all individuals are found above ground during one search of a plot, even under favorable conditions. Two different methods have been used to assess the proportion active at one time. The first was to remove all *P. jordani* repeatedly from replicated plots and to follow the decline in numbers obtained. On one plot, successive removals yielded 31, 30, 13, 6, 13, and 4 specimens; on the other, the numbers were 63, 42, 17, 20, 24, 20, and 8. From these numbers it can be calculated that the average rate of removal was about 0.25 per search, and it was concluded that an average of 25 percent of the individuals present were observed each night (Hairston, unpublished data). The second method was carried out by Nishikawa (1985b). She marked specimens subdermally with fluorescent dye and searched the plots on which they were located over a full season of activity. From recapture data, she estimated the populations of both *P. jordani* and *P. glutinosus*. Knowing the frequency with which each marked specimen was seen, she estimated that each *P. glutinosus* was seen on 24.3 percent of all nights sampled and that each *P. jordani* was seen on 17.9 percent of all nights sampled. The two methods are thus in reasonably good agreement.

The longest run of quantitative counts of any terrestrial salamander species is for *P. jordani* in North Carolina (Hairston 1983b). One set covering five years is for populations in the Great Smoky Mountains (Fig. 2.2); the second is from the Balsam Mountains and runs for eight years (Fig. 2.3). In the same two habitats, *P. glutinosus* populations are shown in Figure 2.4. The differences among the counts are caused by the influence of weather on above-ground activity. Even so, the only dramatic differences came in the autumn, when the observations had to be made after the first cold wave. The aberrant points do not reflect changes in the numbers actually present in the area.

In comparing these graphs, and those that follow, with the population histories presented later in this chapter, it should be kept in mind that nearly all of the salamander data are plotted on arithmetic scales, whereas most of the other data are plotted on logarithmic scales. The effect of the logarithmic scales is to reduce the visual differences among successive means. In other words, the graphs look as though the populations varied less over time than they actually did. The practice of using logarithmic scales is justified in those cases, because the concern is frequently with the proportional changes in

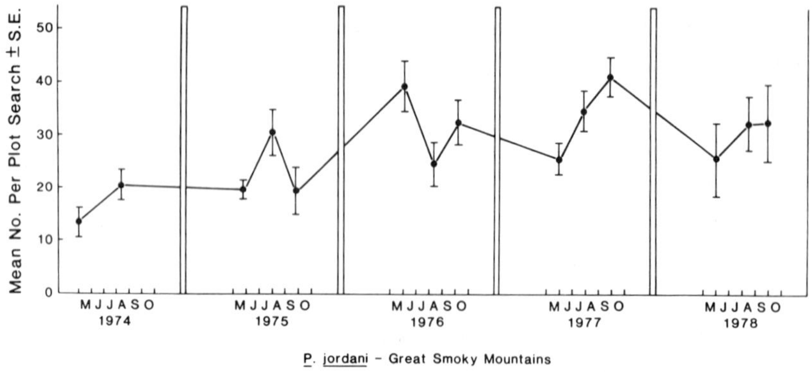

Figure 2.2. Average population history of *P. jordani* during five years on three plots in the Great Smoky Mountains, North Carolina. Plot sizes as in Figure 2.1.

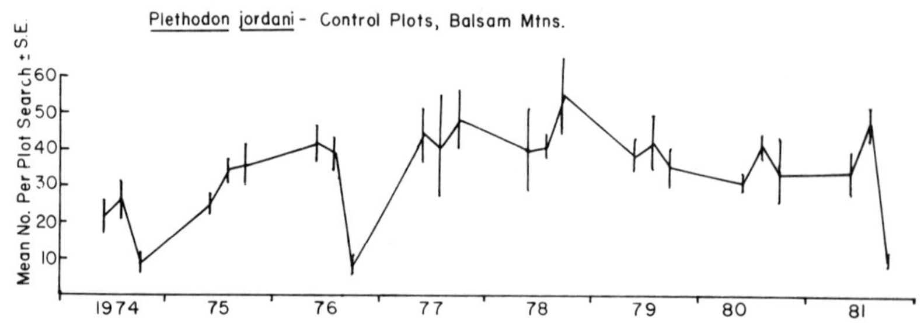

Figure 2.3. Average population history of *P. jordani* during eight years on three plots in the Balsam Mountains, North Carolina. Plot sizes as in Figure 2.1. (From Hairston 1983b.)

population densities, and logarithmic scales are designed to show changes in proportions. Nevertheless, the caution about the visual impression of stability should be remembered.

Thus, the *Plethodon* populations are remarkably stable and demonstrate that assumptions of equilibrium populations in those species appear to be justified. Recent discussions in the literature have pointed out the distinctions between constancy, stability, and equilibrium (Connell & Sousa 1983; Schoener 1985b). Their actual analyses of the literature, however, involved only estimates of constancy, an excessively stringent test of the more interesting equilibrium and stability. In testing for constancy (actually calculating the standard deviation of the log of successive "censuses"), those authors cor-

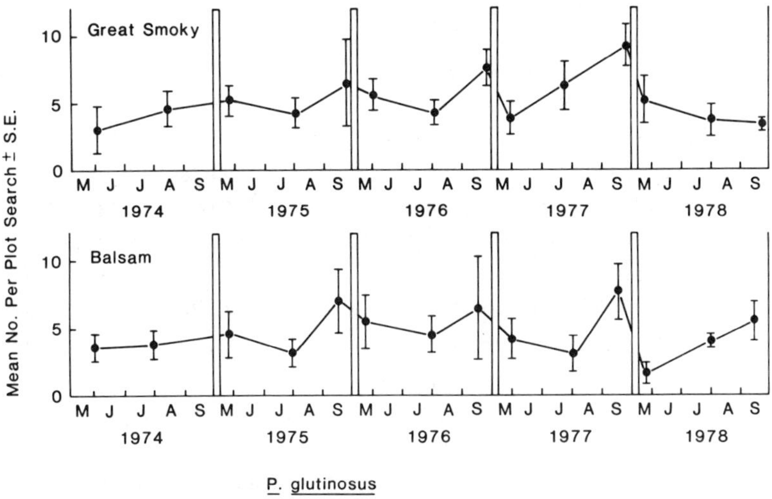

Figure 2.4. Average population histories of *P. glutinosus* during five years on three plots each in (top) the Great Smoky Mountains and (bottom) the Balsam Mountains, North Carolina. Plots are the same as in Figures 2.2 and 2.3.

rectly called attention to the effect of long-lived individuals in reducing the variation in estimates of population abundance. That is because the same individuals would be counted repeatedly. They recommended that successive censuses be used only when there has been a complete turnover of the population (Connell & Sousa 1983) or, more practically, a 95 percent turnover (Schoener 1985b). In the case of these salamanders, there would have to be an interval of more than fifteen years between censuses for *P. jordani*, with its 0.81 survival per year (Hairston 1983b), to meet even Schoener's less rigorous requirement. Thus, to assess stability in salamander populations, there must be recourse to some method in addition to the test for constancy of population size. Connell and Sousa suggested perturbation experiments or testing for constancy of age structure. The latter was done routinely in my salamander studies. In the field, three size classes can be recognized with consistent accuracy. They are the individuals in their first year after hatching or metamorphosis, those in their second year, and a large group of larger individuals consisting of a number of year classes. Large year-to-year changes in the relative abundance of the two younger age classes would mean that the population could not be considered stable in the sense that is usually implied in ecological models. Conversely, if any changes in the age structure are minor, it would mean that reproduction and survival are very similar in successive years, and it can be concluded that the population is in stable equilibrium, at least over the period of the study.

On unmanipulated control plots in the Great Smoky Mountains, *P. jordani* had only minor changes in age distribution over the five years of the experiment (Fig. 2.5). One-year-old individuals were less abundant than two-year-olds because the younger specimens do not emerge from beneath the leaf litter in numbers until late in the summer, and the data were collected over the entire season of salamander activity (May–September). Observations were made by searching marked plots at night for salamanders active on the surface of the ground or climbing on the vegetation. Cover objects were not disturbed, and hence the youngest specimens were underrepresented early in the season.

The longer series of observations on *P. jordani* in the Balsam Mountains shows much the same pattern (Fig. 2.6). Changes in age structure over eight years were small. Comparing these data with Figure 2.5, it can be seen that the proportion of one-year-olds was consistently higher in the population in the Balsam Mountains. The two populations show repeatable differences in age structure, but each shows clear evidence of population stability. Reasons for the greater proportion of one-year-olds in the Balsam Mountains have not been discovered, but it is tempting to relate the difference to the fact that interspecific competition with *P. glutinosus* is intense in the Smokies and weak in the Balsams (Hairston 1980a; see Chapter 5).

The age structures of the two *P. glutinosus* populations (Fig. 2.7) also changed little. They appear to be more variable than the age structures of *P. jordani*, but that is attributable to the smaller sample sizes. *P. jordani* is five or six times as abundant as *P. glutinosus*.

Desmognathus

Unlike *Plethodon*, most species of this genus have aquatic larvae and might be expected to show less regularity in population history. Two studies can be cited as giving information on the stability of *Desmognathus* populations. Tilley (1980) recorded the results of mark–release–recapture analyses of two populations of *D. ochrophaeus* over a period of seven years. Figures 2.8 and 2.9 show the year-to-year variations in his estimates of the numbers of adults present. The sexes are treated separately, but the sex ratio did not vary significantly from 1 : 1. During the last three years, Tilley removed adult males from one of his two sites, and the cumulative numbers have been added to the graph. It is evident that in neither population was there any evidence of systematic variation in numbers during the seven years. He did not publish

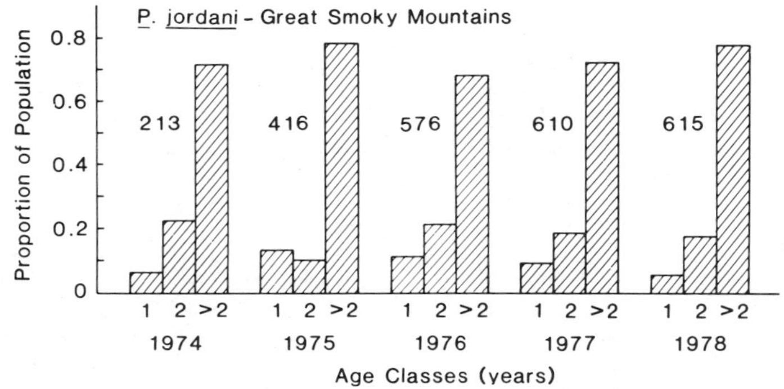

Figure 2.5. Age distribution of *P. jordani* during five successive years in the Great Smoky Mountains. Numbers above histograms give sample sizes.

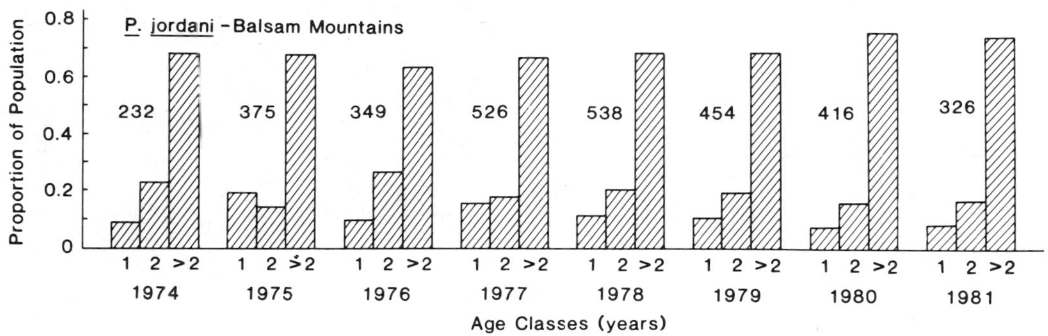

Figure 2.6. Age distribution of *P. jordani* during eight successive years in the Balsam Mountains. Numbers above histograms give sample sizes.

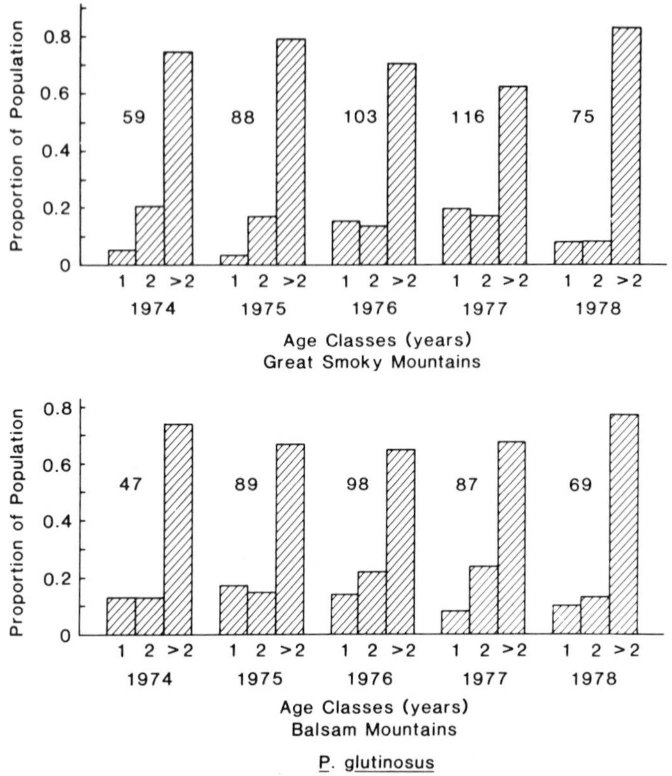

Figure 2.7. Age distribution of *P. glutinosus* during five successive years in two different mountain ranges. Numbers above histograms give sample sizes.

data on immature specimens, but the history of the adult population makes it unlikely that there was any important variation in the number of juveniles present.

My study of *Desmognathus* populations ran for a shorter time, less than three years for comparable data. The habitats of the two studies were markedly different. Tilley worked on populations virtually confined to small, wet rockfaces (38 m^2 for the Whiteside population, and 10 m^2 for Dry Falls). My populations were on forest floor, within 7 m of streams, and the plots covered 140 m^2 (Hairston 1986). For *D. ochrophaeus*, the history is shown in Figure 2.10. The numbers include all animals that had transformed, but, as in the case of the *Plethodon* populations described earlier, the figures represent samples, as the observations were made by counting the individuals visible on

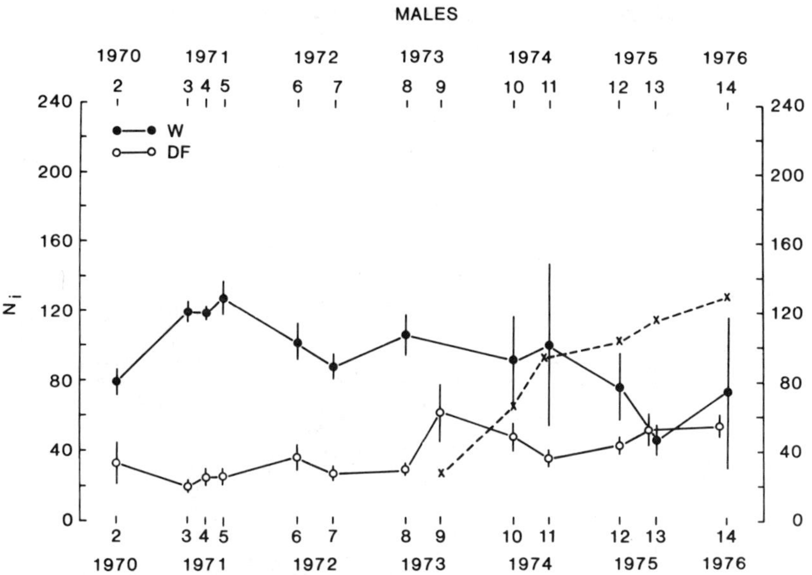

Figure 2.8. Estimated numbers (N_i) of sexable male *D. ochrophaeus* present at each sampling period (i) and their standard errors, for Whiteside (W) and Dry Falls (DF). Estimates with standard errors >100 are omitted because they may be biased. Crosses denote cumulative numbers of males removed from the Whiteside population between August 1973 and June 1976. (From Tilley 1980.)

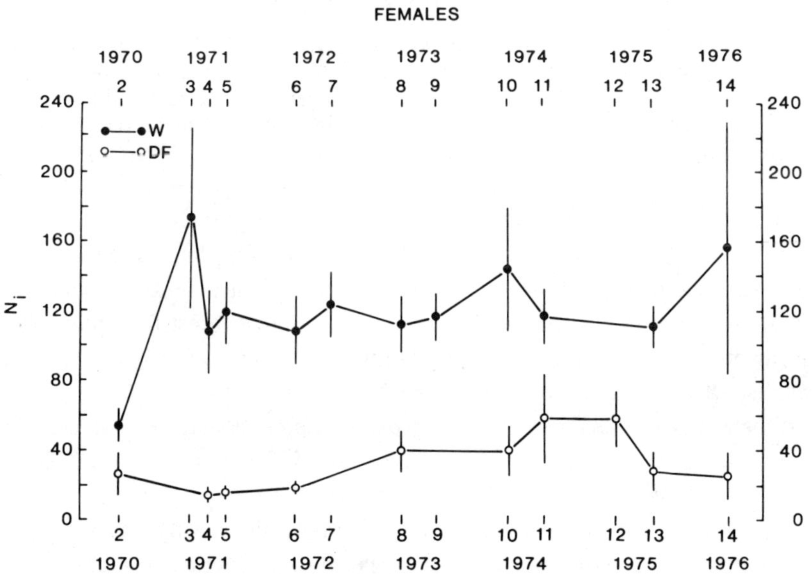

Figure 2.9. Estimated numbers of sexable females and their standard errors, for Whiteside and Dry Falls. Notation as in Figure 2.8. (From Tilley 1980.)

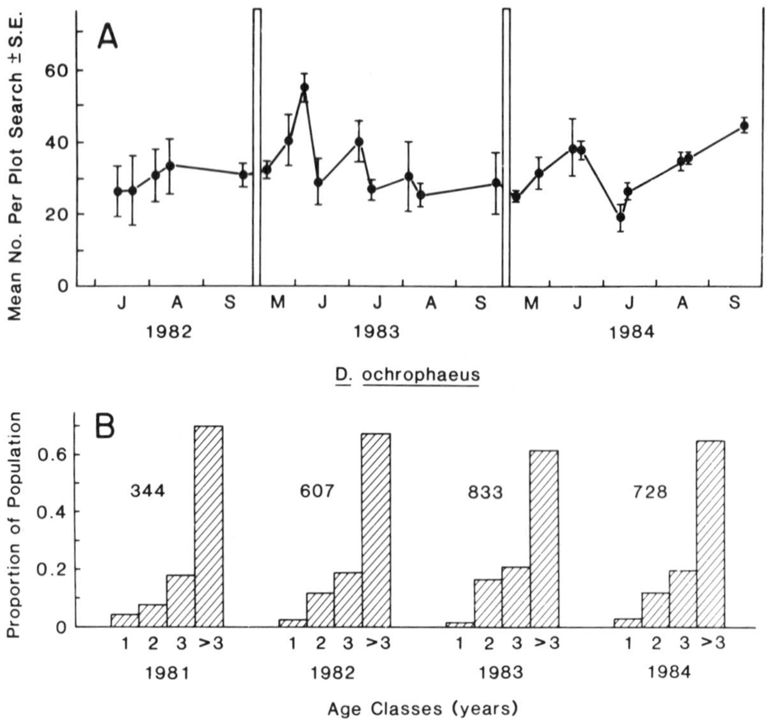

Figure 2.10. (A) Changes in density of *D. ochrophaeus* for three years on control plots (means and 1 standard error). (B) Changes in age distribution on the same plots (in 1981, plots were larger and were searched less intensively than in the other three years). Numbers above histograms give sample sizes.

the surface of the ground or climbing. On the basis of repeated removals from experimental plots over a short time, I estimate that less than 10 percent of the individuals present are visible during a search of these forest-floor plots. The mean number seen per plot visit was 27 for control plots, of which 17.63 were adults. Because each plot was searched three times at each visit, it can be estimated that 21 percent of the *D. ochrophaeus* present were seen. Hence, the estimated density was 83.51 adults per plot (0.6 per m^2). From Tilley's statements it can be estimated that on the rockfaces, 20–25 percent of the adults were in sight at each visit. In those habitats, the density was 5.8 adults per square meter on one rockface and 6.9 per square meter on the other. These are ten times the estimate for the forest floor, despite the apparently greater abundance of food in the latter location. As can be seen from comparing the

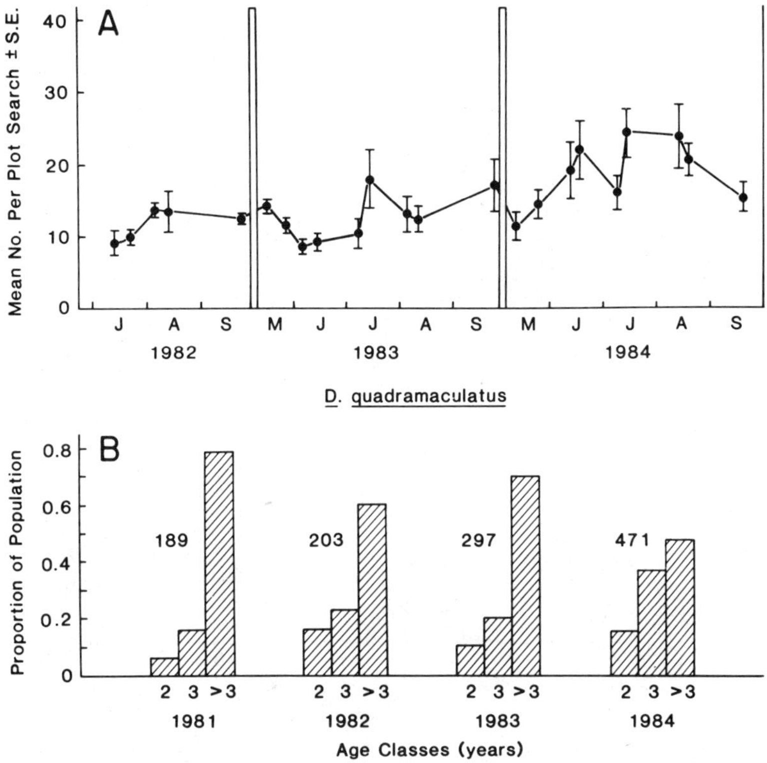

Figure 2.11. Changes in (A) population density and (B) age distribution of *D. quadramaculatus*. Details as given for Figure 2.10.

three figures, the differences in density were not related to the stability of the three populations, which was impressive in each case. These observations are discussed further in Chapter 6. As with *Plethodon*, the age distribution of *D. ochrophaeus* changed little over the four years of my study (Fig. 2.10). The newly metamorphosed individuals are very small and are easily overlooked, but the overall picture confirms the stability of the population.

Two other species of *Desmognathus* were abundant enough on these plots to allow me to make accurate estimates of their populations during the same three seasons. *D. quadramaculatus* is a large, mostly aquatic species with a larval period of at least two years. Figure 2.11 shows how little variation there was in the numbers seen. There appeared to be a gradual increase in number over the three seasons, but the difference, while statistically significant, was minor. *D. monticola*, a streambank species intermediate in size between *D. ochrophaeus* and *D. quadramaculatus*, showed fluctuations in mean numbers

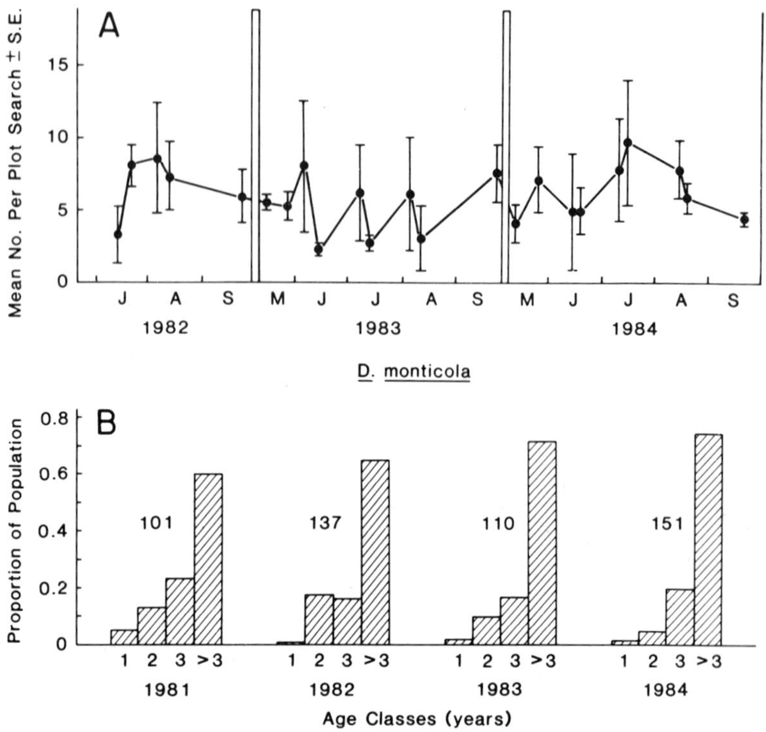

Figure 2.12. Changes in (A) population density and (B) age structure of *D. monticola*. Details as given for Figure 2.10.

that were related to weather conditions, with no long-term trends (Fig. 2.12). For both *D. quadramaculatus* and *D. monticola*, the successive age distributions confirm that reproduction and survival figures remained quite similar over the four years of the study. These studies on *Desmognathus* reveal that plethodontid salamanders with amphibious life histories have populations that are as stable as the populations of those with completely terrestrial life histories.

Pond-breeding species

The population histories of salamander species that breed in ponds are complicated by the fact that it has not been possible to estimate numbers at times other than the breeding season. The effect of this difficulty is to make the data dependent on conditions affecting the proportion of adults that make the annual migration. As is mentioned in Chapter 3, the weather during the season is crucial in determining that proportion. In some species, the duration of the period over which migration can safely be undertaken is short. The example of *Ambystoma maculatum* has been given. In some areas, it is caught between weather that is too cold and weather that is warm enough for the activity of its snake predators (C. R. Shoop, pers. commun.). In some seasons, spring arrives rapidly, and many individuals are known to break off their migration and return toward their forest home ranges. These factors make for major annual variations in the numbers of adults arriving at the breeding ponds.

The magnitudes of some of these annual variations are presented in Chapter 3 (Table 3.3). The nearly eightfold difference in numbers arriving to breed in different years (12.6-fold for females) makes direct assessment of population stability impossible, because the known longevity of adults precludes the interpretation that the fluctuations represent real changes in total abundance. What is required in such cases is knowledge of age-specific survival and fecundity schedules, so that a life table can be drawn up. Unfortunately, the methods for estimating age-specific survival and fecundity often require the assumption of population stability. That would, of course, negate the use of the life table in assessing population trends. In those cases where the life table can be made independent of the unwanted assumption, data are needed over a long span of years, in order to average out the immediate effects of weather on reproductive success in different years. The life table of *Ambystoma talpoideum* is given in Chapter 3. The calculated net reproductive rate of 1.032 shows that the population was essentially stable during the five years of Semlitsch's study (Semlitsch 1985a, 1987, pers. commun.).

Douglas E. Gill has obtained data on the numbers of two other species of *Ambystoma* incidental to his long-term study of newt populations in Virginia (see next paragraph). The data for *A. maculatum* are less variable than are those for other species or for other populations of *A. maculatum*, in that the numbers arriving at his ponds varied little from year to year. *Ambystoma jeffersonianum* in the same ponds had a population history that was more typical, with at least fivefold differences between years in the numbers arriving at the same pond.

Gill, Berven, and Mock (1983) reported the results of six years' observations on a "metapopulation" of *Notophthalmus viridescens*. The metapopulation consisted of the newts breeding in a group of ponds in the mountains of Virginia. The population histories for five of the ponds are shown in Figure 2.13. Two of the ponds experienced an increase in the number of adults entering each year; the other three had virtually constant populations. The

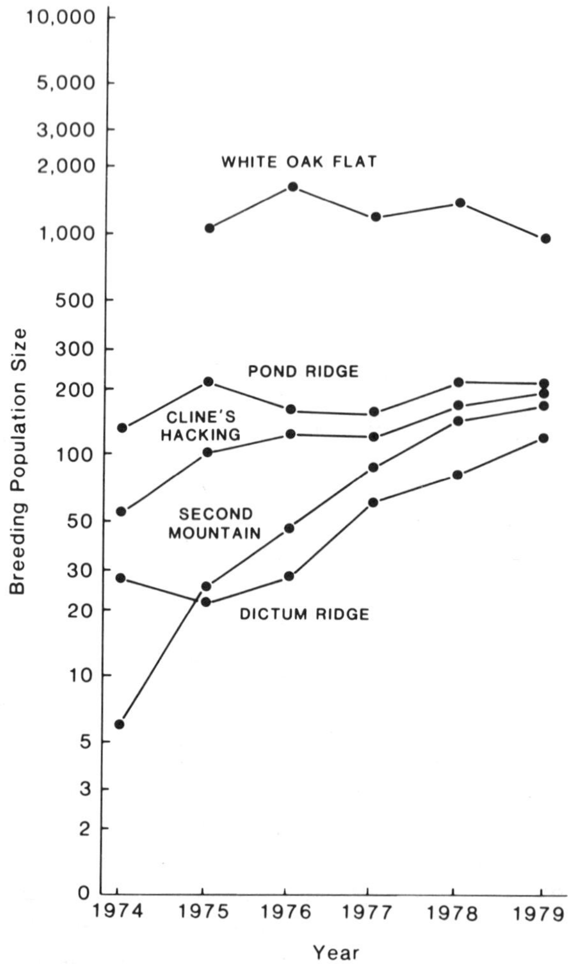

Figure 2.13. Breeding population sizes of red-spotted newts in five neighboring ponds in Virginia over a period of eight years. (From Gill et al. 1983.)

appearance of stability was enhanced by using a logarithmic plot of numbers, as discussed earlier. Nevertheless, the whole metapopulation changed little during the period of Gill's study (1978, 1979, 1985), and moreover the variation between ponds became less over the years. That is the more surprising, in view of his observation that production of metamorphosed juveniles was highly irregular in all of the ponds that he observed. For brief periods of one or two years at a time, each pond population produced large numbers of juveniles, and only in those years was production sufficient to exceed losses from estimated mortality (Gill et al. 1983). His revised estimates of the rate of survival (Gill 1985) show that *Notophthalmus*, like many other species of salamanders, has an excellent chance for a long life (see the next section), and the longevity no doubt accounts for the stability of the populations in those ponds where a marked decline would otherwise be expected.

Longevity

Survival in nature

Three methods may be used to compare the longevities of individuals of different species. They are to determine the ages of individuals, to follow recaptures of marked individuals, and to use an indirect method that assumes a constant rate of survival. The annual proportion surviving has been estimated for a number of species. The most meaningful comparison in this case is of adult animals, as the survival of immature individuals depends largely on the mode of life history followed. The first attempt to estimate annual survival rates was that of Organ (1961). He used the number of lobes on the testes of male *Desmognathus*, on the assumption, following Humphrey (1922), that a lobe is added for each two years of sexual maturity. By making large collections and recording the number of males in each lobe category, he arrived at data that permit the following estimates: *D. quadramaculatus* had approximately equal survival rates over each successive biennium: 0.184, 0.148, and 0.252. The weighted mean of these is 0.180, from which the annual survival of 0.424 can be calculated as e^{-a}, where a is the annual death rate. It is obtained from

$$l_x = e^{-ax} \tag{2.1}$$

where l_x is the proportion surviving for x years. The other four species had decreasing survival rates in the last biennium or in the last two. *D. monticola* had successive survivals of 0.214, 0.266, 0.274, and 0.167. The mean is 0.230, and the annual survival 0.482. If we ignore the last biennial survival as perhaps being explicable on the basis of insufficient room on the spermatic cord for the indefinite addition of lobes (Tilley 1977), the annual survival was 0.500. The other three species in Organ's study had decreasing survivals during the last two biennia. Based on weighted means over all ages, *D. fuscus* had an annual survival of 0.557. Over the first two biennia only, the rate was 0.583. For *D. ochrophaeus*, the respective rates are 0.656 and 0.665, and for *D. wrighti* they are 0.687 and 0.786. Whichever method of calculation is used, there is a clear negative relationship between size and annual survival.

Danstedt (1975) used the same method to estimate the survival of *D. fuscus* in six distinct populations in Maryland. His estimates range from 0.238 to 0.420 per year for males, and from 0.268 to 0.426 for females. Thus, even the high survival rates in his populations are somewhat lower than would be indicated by either method of estimating survival from Organ's data for *D. fuscus*.

Salamanders are known to regenerate parts, such as toes or even whole limbs, that have been removed in mark–recapture experiments. The process normally is slow, however, and in several studies toe clipping has been used to follow the history of individual salamanders. No evidence has been found that such marking has adverse effects, and the technique has been applied in important population studies to *Triturus vulgaris* (Bell 1977), *Taricha rivularis* (Twitty 1966), *Notophthalmus viridescens* (Healy 1970, 1974), *Ambystoma maculatum* (Husting 1965; Shoop 1974), *A. tigrinum* (Semlitsch 1983a), and *A. talpoideum* (Semlitsch 1985a, 1987).

Other investigators have taken advantage of the fact that in some species each salamander has a unique set of color markings, which it retains throughout life. This characteristic has been used in studying the populations of *Notophthalmus viridescens* (Gill 1978, 1985), *Desmognathus ochrophaeus* (Tilley 1977, 1980), and *Ambystoma opacum* and *A. maculatum* (Stenhouse 1984).

Tilley (1980) used the characteristic dorsal color patterns of *D. ochrophaeus* to follow individals over several years. Adults in the two populations with which he worked were quite different in annual survival, even though the ecological situations were similar: wet, vertical rockfaces. At the high-altitude site (1,448 m on Whiteside Mountain), he estimated survival as 0.743 per year. At the lower site (1,024 m at Dry Falls), the annual survival was 0.626. His estimates thus bracket those of Organ for the same species.

The long study of *Notophthalmus viridescens*, the newt of eastern North America, by Gill (1978, 1985) also involved the recognition of individuals by their unique color patterns. Gill had sufficient data to allow him to analyze survival separately for the sexes, for the different years of his study, and for five of the ponds inhabited by the same metapopulation. Survival of adult males had a range from 0.675 to 0.83 per year in different ponds; for adult females, the range was from 0.63 to 0.74 per year. These differences are statistically significant, as are the differences between years and between ponds.

Toe clipping permits ready identification of individual animals and has yielded annual survival rates of impressive magnitude. In a massive study of the red-bellied newt of California (*Taricha rivularis*), Twitty (1966) marked large numbers of animals over a period of thirteen years. Failure to recover an individual does not necessarily mean that it was no longer in existence, and the derived survival rates are therefore minimal. Eleven years after the first group of 2,097 was marked, 791 were recaptured. If survival were constant each year, this 38 percent would be the annual survival raised to the eleventh power, giving an estimate of 0.91 per year. Groups marked in other years gave similar results, and the annual survival in those cases can be calculated to range from 0.89 to 0.92, for a mean length of life of around ten years.

A third salamandrid was followed long enough by toe clipping to obtain valid survival estimates: the European smooth newt, *Triturus vulgaris*. This species has the interesting property that metamorphosed but immature individuals have higher survival rates than do the adults. Bell (1977) calculated that survival of adult males was 0.45 per year, and for adult females, 0.55 per year. Immature terrestrial forms had an annual survival that can be calculated as 0.798. It is possible that efts of *Notophthalmus* also have better survival than adults, but no quantitative estimates are available.

Studies of two species of *Ambystoma* have yielded survival rates for the adults. Husting (1965) followed a population of *A. maculatum* for five years, marking individuals with toe clips. He calculated mean survival of males as 0.79 per year, and for females, 0.63 per year. He had low proportions returning in any given year, a fact that Gill (1985) attributed to frequent avoidance of the fence by the migrating adults, a trait that he referred to as "trespassing." Gill was able to assess the frequency with which it took place and found that he obtained much higher survival estimates than he had obtained (Gill 1978) before he made the correction. Unfortunately, it is not possible to make a correction of Husting's calculations, if such a correction is needed.

Semlitsch (1985a, 1987, pers. commun.) had a fence that was virtually trespass-proof in his study of *Ambystoma talpoideum*. Failure of his marked animals to return to the pond to breed was therefore genuine. Unfortunately, one of his five years of observation was disastrous, with very few migrating adults, and virtually no production of young. If we ignore the data from that year, survival of adults can be calculated to be 0.566 per year.

In my study of populations of the completely terrestrial *Plethodon jordani*, I did not mark individual salamanders (Hairston 1983b). Survival of adults was estimated by being able to identify the first-year class of the group of large animals composed of all ages beyond two years. Figure 3.7 shows that there must be a number of age classes in this one size class. I assumed that once adulthood had been reached, survival was constant. In studies in which it has been possible to obtain year-by-year survival rates, this assumption has been found to be approximately correct (e.g., Gill 1985). The annual survival is therefore the value that would permit the accumulation of enough age classes to account for the relative number of individuals in the first-year class and those older than that. The annual survival of adult *P. jordani* was calculated to be 0.81.

The mean generation time

The second and more comprehensive approach to longevity is to calculate the mean generation time, a statistic that requires the construction of a life table. It is described as the mean age of giving birth, or the mean age of egg laying. Mathematically, it is expressed in the equation

$$R_0 = e^{rT} \tag{2.2}$$

where R_0 is the net reproductive rate, r is the intrinsic rate of natural increase, and T is the mean generation time (Slobodkin 1962). The equation applies only to populations that are changing, as $r = 0$ in stationary populations. In such populations, the following equation is used:

$$T = \Sigma x l_x m_x \tag{2.3}$$

Here, l_x is the proportion of females surviving from birth (or egg laying) to age x, and m_x is the mean number of female offspring produced per female during the time interval of which x is the midpoint. Five studies, covering a total of eight species, have been complete enough to permit the construction of life tables. The first complete life tables of salamanders were those assembled by Organ (1961) for five species of *Desmognathus*. He made direct observations of clutch sizes, but used the number of adult females to calculate total egg production, from which he could calculate the survival of larvae, as his collections gave him the number in that stage of the life history. Survival of immature animals was obtained from the numbers in successive size classes. The ages of adult males, which could not be separated by size, were obtained from the number of lobes on the testes, as described earlier. Female survival was derived from knowledge of the frequency of production of clutches (alternate years), the sex ratio of adults, and the assumption that females not engaged in reproduction had the same survival as males of the same species. Although survival rates for adult males were different between the five species, the mean generation times were very similar: 5.35 years for *D. quadramaculatus*, 5.56 years for *D. monticola*, 5.14 years for *D. fuscus*, 5.46 years for *D. ochrophaeus*, and 5.31 years for *D. wrighti*.

Nearly twenty years later, Tilley (1980) published life tables for his two populations of *D. ochrophaeus* living on rockfaces. His tables differ from those of Organ in that he found annual breeding and lower survival to maturity. The combination yielded, for one of his populations, a mean generation time of 5.6 years, a similarity to Organ's finding that is clearly fortuitous. His other population was estimated to have a mean duration of 7.5 years. The geographic area (230 km south of Organ's site) and the ecological setting (rockfaces) may account for the differences (Hairston & Tilley, unpublished data).

Bell's work (1977) on *Triturus vulgaris* resulted in a complete life table. Unlike the other examples that have been produced, this life table revealed a declining total and thus agreed with annual population estimates. The net reproductive rate was only 0.328, and the population in Bell's best-studied pond declined from a calculated 999 to 293.5 in three years. Bell calculated the mean generation time as approximately 7.6 years. The pond was not atypical; the other seven ponds for which data are available for two years all had declines in newt populations, and the predicted decrease was less than those observed.

The shortest mean generation time that has been calculated for any salamander is that for *Ambystoma talpoideum* (Table 3.5 in Chapter 3, data of Semlitsch). The calculation gives 2.6 years, and its brevity is related to the short immature period of one year. Short though it is, in comparison with other salamanders, a generation of that length exceeds the period of seasonal fluctuations in weather, and the lives of the animals extend over several years. This means that differences between years, though they affect the number of adults reproducing in each year, have much less serious implications for population fluctuations than would be the case with species whose generations lasted one year or less.

The construction of the life table for *Plethodon jordani* is described in Chapter 3. Like most of the others, it shows a nearly unchanging population, and like most of the others, there is an element of circularity. As already demonstrated, however, the long-term stability of the population was established in an independent manner. The calculated mean generation time of *P. jordani* is 9.79 years. That is the longest yet determined, but it holds that position only because no life table has been constructed for *P. glutinosus* or for *Taricha rivularis*. *P. glutinosus* has an immature period a year longer than *P. jordani* and can scarcely fail to have a longer generation time; *T. rivularis*, with its mean length of adult life of ten years, almost certainly has longer generations.

Captive specimens

The potential for survival has often been determined by keeping animals in captivity, where predation and starvation, at least, can be prevented. More

Table 2.2. *Known longevity (in years) of captive salamanders*

Family and species	Max. age	Authority
Cryptobranchidae (*Andrias japonicus*)	55	Goin et al. (1978)
Amphiumidae (*Amphiuma means*)	27	Goin et al. (1978)
Proteidae (*Proteus anguinus*)	15	Goin et al. (1978)
(*Necturus lewisi*)	18	Martof et al. (1980)
(*Necturus punctatus*)	10	Martof et al. (1980)
Sirenidae (*Siren lacertina*)	25	Goin et al. (1978)
Ambystomatidae (*Ambystoma maculatum*)	25	Goin et al. (1978)
(*Ambystoma tigrinum*)	11	Bishop (1941)
Salamandridae (*Cynops pyrrhogaster*)	25	Goin et al. (1978)
(*Triturus cristatus*)		
Male	25	Frazer (1983)
Female	17	Frazer (1983)
(*Triturus vulgaris*)	20	Frazer (1983)
(*Triturus helveticus*)	12	Frazer (1983)

than most biologists, herpetologists are happiest when keeping the objects of their interest nearby. The result of this enthusiasm is that there are records for many species of salamanders extending over surprising lengths of time. Many of these records are given by Goin, Goin, and Zug (1978), and more are found in other herpetological writings. For obvious reasons, what we have are the maximum known lengths of life, and there is no means of obtaining average survival in captivity. The champion is the giant cryptobranchid, *Andrias japonicus*, which has lived for fifty-five years after capture. Table 2.2 gives the records for the other species with unusual longevity. It is clear that salamanders are capable of living for many years, and like a number of species of reptiles, they do not appear to have finite life spans. Most likely, under natural conditions, they live until they are killed.

Assessment of stability in other taxonomic groups

For comparison with the salamander populations described earlier, I have selected several examples of populations that have been followed long enough to meet the criterion of Connell and Sousa (1983), in that they were followed long enough to cover the life spans of all individuals. It was also important that a number of "censuses" had been taken, because Connell found that studies in which only two estimates were made had significantly lower variances than did those with more counts. The studies were also selected to represent organisms in freshwater and terrestrial habitats. There

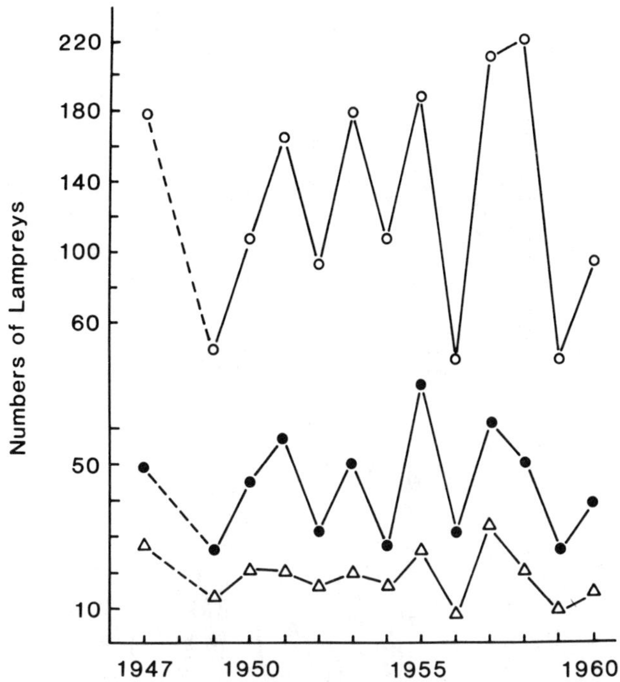

Figure 2.14. Fluctuations in annual spawning populations as measured by total numbers (top), peak catch (middle), and average catch (bottom). (From Hardisty 1961.)

are no marine salamanders, and marine species have been omitted, as have all sessile organisms. No attempt was made to select either species with highly variable populations or those with relatively constant populations.

Freshwater species

The first example of a study on an aquatic species is Hardisty's fourteen-year analysis of an isolated population of the brook lamprey, *Lampetra planeri*, in an English stream (Hardisty 1961). The year-to-year variation in the spawning population was as great as fourfold (Fig. 2.14). It should be remembered that lampreys have prolonged life cycles. This species spawns and dies at six or seven years of age, and the study therefore covered two turnover periods. An examination of the figure shows how the variance could be lower for two censuses than for a larger number, as pointed out by Connell and Sousa (1983).

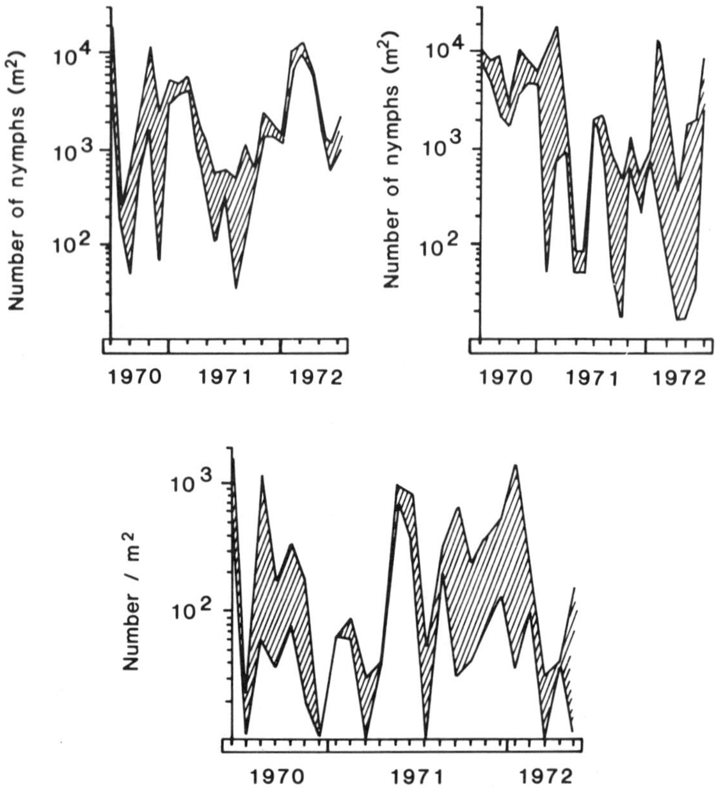

Figure 2.15. Population estimates for nymphs of a mayfly species in three locations in Idaho. (From Newell & Minshall 1978.)

Insect populations have come in for a great deal of attention because many species are of economic importance. Several such examples will be given later, and I have selected one aquatic species to show that forest insects are not exceptional in the magnitude of population fluctuations. Newell and Minshall (1978) followed a population of the mayfly *Tricorythodes minutus* over a three-year period in two streams in Idaho. One of the streams was spring-fed and virtually constant in temperature; the other varied seasonally from 0 to 29 °C. In Figure 2.15, the upper two graphs show the populations in two parts of the stream with a constant temperature, where the species is multivoltine. The lower graph is for the population in the stream with a variable temperature, where it is bivoltine. There are considerable differences between samples at each location, but the data leave no doubt that each population varied by as much as 100-fold during the three years. That is well beyond the level of fluctuation of any salamander population.

The next example is my study of the protozoan *Paramecium* (Hairston 1967). The population was followed for twenty-one months, or approximately twenty-one turnover periods, by Schoener's method of calculation (Schoener 1985b). During that time, the numbers fluctuated through 2.5 orders of magnitude (Fig. 2.16), a far wider range than for any of the salamander populations described earlier in this chapter. Despite the clear lack of a constant population, there was good evidence that it was tracking the carrying capacity of the environment and that it was responding in a density-dependent way. Thus, although it would not have met the criteria of either Connell and Sousa (1983) or Schoener (1985b) for constancy, it would be difficult to deny that in some sense it was in equilibrium.

Terrestrial animals

It is frequently stated that insect populations show great irregularities in density, a feature that has been called "density-vague population dynamics" (Strong 1983). Inasmuch as insects form such a large proportion of the fauna, it is appropriate to include some populations for comparison with salamanders. Varley (1949) summarized a very large amount of information on forest insects in Germany. The data cover sixty years for four different species (Fig. 2.17). Most of the populations fluctuated over at least two orders of magnitude, although there were periods of as much as ten years when some of the populations remained within much narrower limits. Although not avail able for these data, the variability of individual counts may be judged for *Bupalus* from the graph provided by Klomp (1966) for the same species in the Netherlands (Fig. 2.18). An unusually long run of population density estimates has been provided by Baltensweiler (1968) for the moth *Zeiraphera griseana*. He also gave the range of means for samplings in different sectors of the Swiss canton Graubunden. The population went through two major fluctuations of three to four orders of magnitude during eighteen years.

Terrestrial vertebrates provide important comparisons with salamanders because of their contribution to community theory. Lizards, small mammals, and especially birds have provided the data on which much thinking has been based. Schoener (1985b) presented data demonstrating that many lizard populations varied little between censuses taken at least as far apart as the maximum duration of a generation. Most of the studies cited involved only two censuses, but the study of *Uta stansburiana* by Tinkle (1967) covered adult

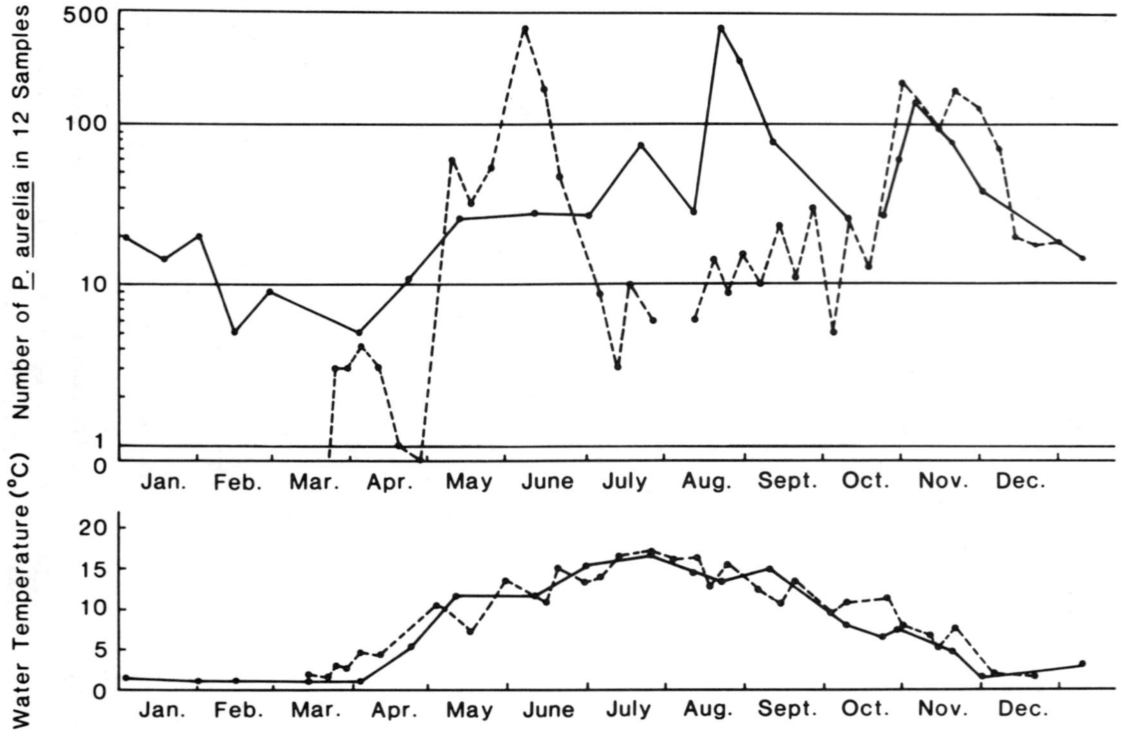

Figure 2.16. Natural changes in the population of *Paramecium aurelia* over a period of two years. The broken line is for 1963; the solid line is for 1964. Gaps represent periods of experimental manipulations. Temperatures are averages for the water in twelve samples on each date. (From Hairston 1967.)

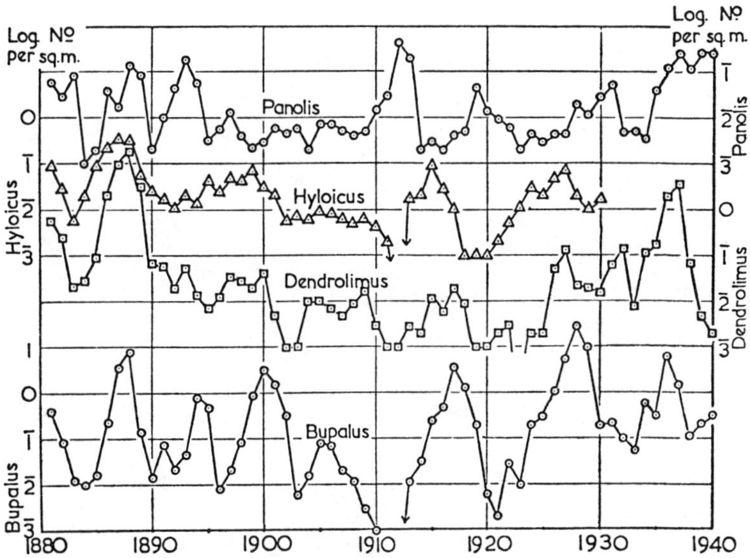

Figure 2.17. Results of sixty successive winter census counts of moth pupae (or, in the case of *Dendrolimus*, hibernating larvae) per square meter of forest floor at Letzlingen, Germany, replotted on a logarithmic scale. (From Varley 1949.)

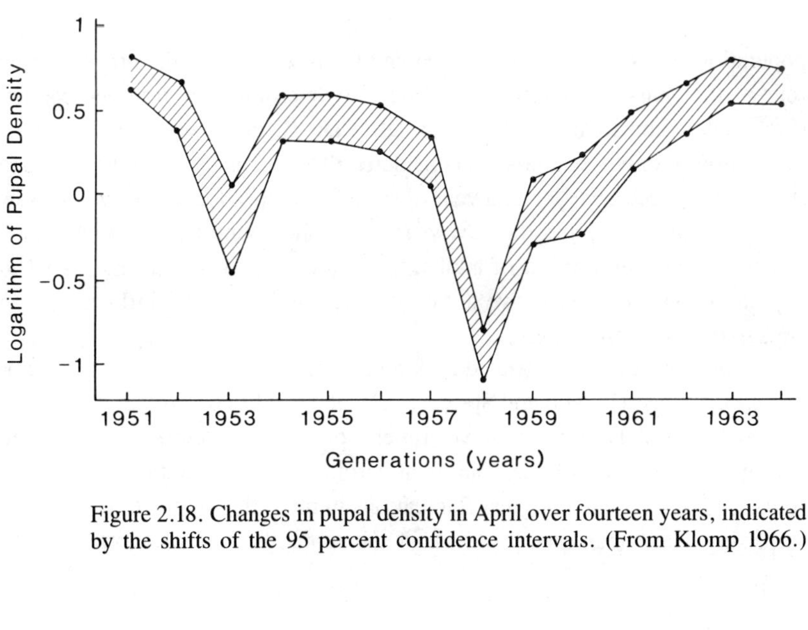

Figure 2.18. Changes in pupal density in April over fourteen years, indicated by the shifts of the 95 percent confidence intervals. (From Klomp 1966.)

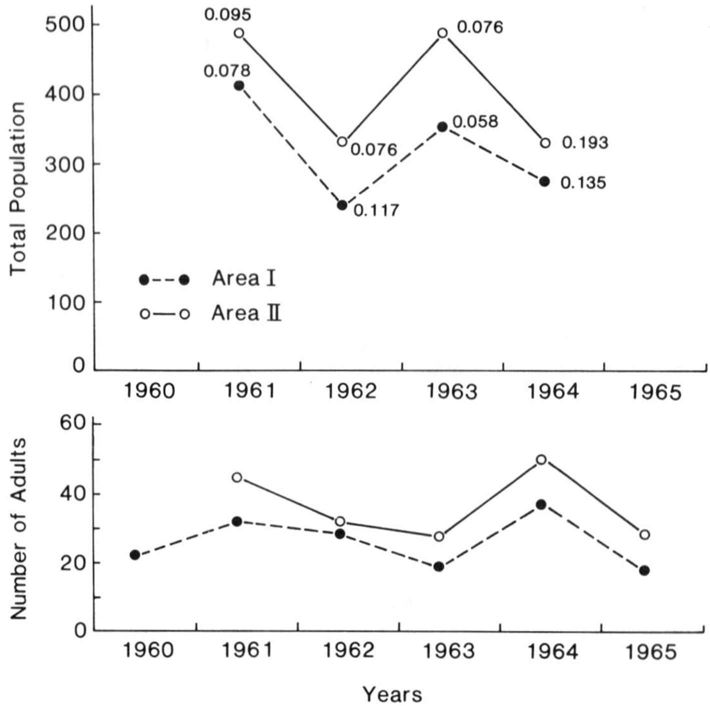

Figure 2.19. Censuses of *U. stansburiana* in western Texas. (Data from Tinkle 1967.)

populations for six years on one area and for five years on the other. During four of the years, the entire population was determined by repeated capture (Fig. 2.19). The total variation was less than twofold, whether the whole population or only the adults are considered. There was more variation in age structure in successive years, a year of successful reproduction showing a low proportion of adults, and the next year (when the young had become adults) showing a high proportion of adults. With a short generation time, the *Uta* population was more sensitive to year-to-year environmental influences than are salamander populations.

Microtine rodents have tended to dominate ecological thinking about mammal populations. Connell and Sousa (1983) cited studies of twelve populations of small mammals with "turnover times" of one year; eight of these were microtines, and two of the four nonmicrotines became extinct during the studies. Given the known great fluctuations in microtine populations, it is not surprising that "mammals" placed in the more variable categories on the

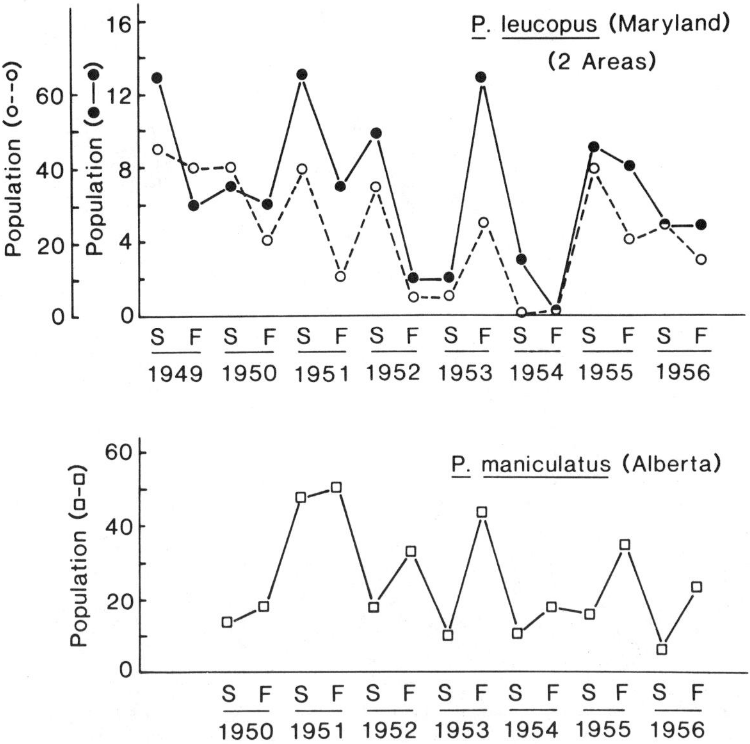

Figure 2.20. Population estimates for two species of *Peromyscus* made twice per year for a number of years. (Data from Terman 1968.)

graph by Connell and Sousa. Other small rodents, such as *Peromyscus*, were underrepresented, and two such populations are shown here (Fig. 2.20). The data are taken from Terman (1968). Using spring data, both species had the same standard deviations of the logarithms of censuses: 0.28, a value that places them below any of the mammals on the Connell and Sousa graph.

Little need be said about the influence of birds on the development of theories of animal communities. From Lack (1947, 1966) and MacArthur (1958, 1972) to Grant (1968, 1983), Diamond (1975, 1982), and others, community theorists have relied heavily on their knowledge of birds. Most bird populations appear to remain relatively constant (Connell & Sousa 1983), despite the example of the great and blue tits near Oxford (Fig. 2.21). One well-known species, however, the red grouse, has populations going through much wider fluctuations (Fig. 2.22). The data included in this figure were assembled from Jenkins, Watson, and Miller (1967), Moss, Watson, and Parr

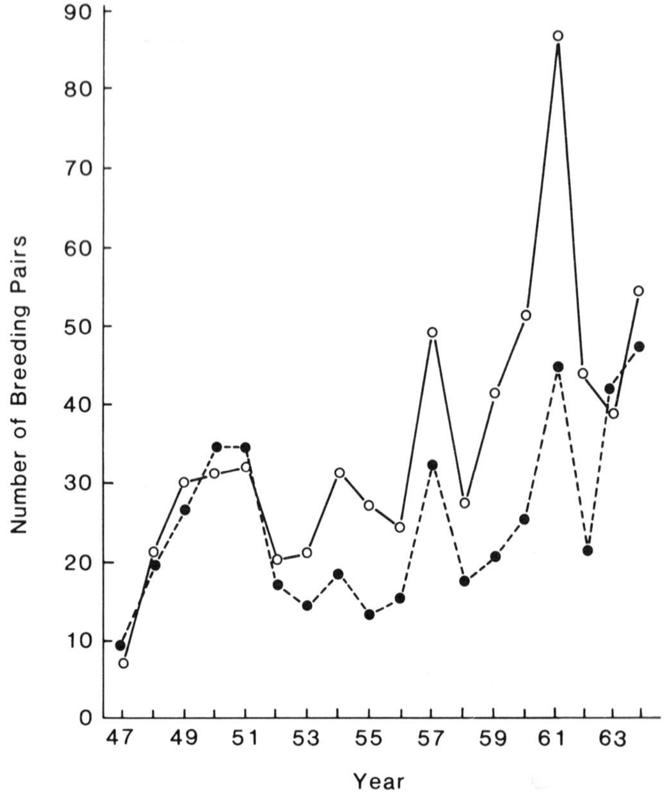

Figure 2.21. Numbers of breeding pairs of great (open circles) and blue tits (filled circles) in Marley Wood, 1947–64. (From Lack 1966.)

(1975), and Moss, Watson, and Rothery (1984). As has been noted before, the apparently cyclic nature of the fluctuations is reminiscent of those for some mammals, such as the snowshoe hare. Cycles have frequently been attributed to predator–prey relationships. In the case of the red grouse, a nematode parasite, *Trichostrongylus tenuis*, has been implicated by field experiments and mathematical models as the driving factor in the cycles (A. Dobson, pers. commun.).

This brief presentation of the population histories of other kinds of species shows clearly that few of them compare favorably with salamanders in the constancy of their populations. Constancy is not a necessary condition for equilibrium, but over a reasonable amount of time even relative constancy implies something about stability. To remain "constant" through periods when other species respond to environmental perturbations at least implies a

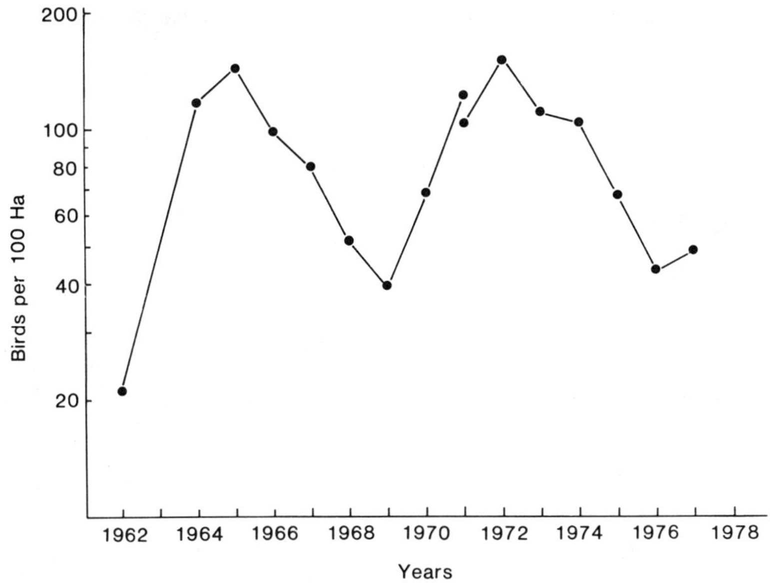

Figure 2.22. Estimated number of red grouse per 100 ha at Kerloch, Scotland. (Data from Jenkins et al. 1967; Moss et al. 1975, 1984.) The gap in 1971 represents a difference between data in two publications.

greater stability on the part of the constant species. No species is immune to all perturbations, and to argue that we must know the magnitude of a perturbation necessary to cause a change in abundance before we can assess its stability is to ignore the relative nature of all such assessments. Small fluctuations over long periods, especially combined with the demonstrated constancy of age structure, show that salamanders are highly appropriate material for testing community theory that relies, implicitly or overtly, on assumptions of equilibrium and stability.

Synthesis

Salamander populations are apparently little affected by stochastic events, unless these events are destructive of the habitat. The exceptions occur with the variations in numbers of adults migrating to breeding ponds, the variations depending on weather over a short time. The remaining species, some populations of which have been followed for as long as eight years, show inconsequential fluctuations in numbers counted on the surface of the ground. These minor fluctuations are caused by the responses of the animals

Table 2.3. *Annual survival of adult salamanders and mean generation times for most of the same species*

Species	Sex	Annual survival	Mean generation time	Source
Desmognathus quadramaculatus	M	0.482	5.35	1
D. monticola	M	0.500	5.56	1
D. fuscus	M	0.583	5.14	1
	M	0.238		2
	M	0.420		2
	F	0.268		2
	F	0.426		2
D. ochrophaeus	M	0.665	5.46	1
	M	0.626	5.6	3
	M	0.743	7.5	3
D. wrighti	M	0.786	5.31	1
Notophthalmus viridescens	M	0.675		4
	M	0.83		4
	F	0.63		4
	F	0.74		4
Taricha rivularis	M	0.91		5
Triturus vulgaris	M	0.45	7.6	6
	F	0.55	7.6	6
Ambystoma maculatum		0.79		7
A. talpoideum		0.566	2.6	8
Plethodon jordani		0.81	9.79	9

Sources: 1, Organ (1961); 2, Danstedt (1975); 3, Tilley (1980); 4, Gill (1985); 5, Twitty (1966); 6, Bell (1977); 7, Husting (1965); 8, Semlitsch (1986a); 9, Hairston (1983b). Two values for the same species and sex by the same author represent different areas.

to ambient weather conditions, as they remain under ground in exceptionally cold or dry periods. Confirmation of the year-to-year regularity in both reproduction and survival comes from the successive age distributions of seven different plethodontid populations followed from four to eight years. The foregoing generalizations apply both to terrestrial species (*Plethodon*) and to those with an aquatic larval stage (*Desmognathus*). Pond-breeding species may show nearly constant numbers of adults arriving each spring, as is the case with *Notophthalmus*, or may be strongly influenced by weather during the normal migrating season, as is true of several species of *Ambystoma*.

Stability of populations is greatly enhanced by increased longevity of adult animals, and many species of salamanders have been found to survive for surprisingly long times. The annual survival rate may be found by any of several methods. If the ages of all individuals can be estimated, the ratios between the numbers at successive ages give the survival rates. Individuals have been followed by marking them, or by recognizing color patterns that are specific to them. Sufficient numbers of records also yield estimates of annual survival. The available information for salamanders is given in Table 2.3. The range of values for adults is from less than 50 percent per year to 91 percent per year. Average lengths of adult life are thus from less than two years to more than ten.

A more general assessment of the population is gained by knowledge of the mean generation time, which incorporates survival through different ages with age-specific reproduction. The available information is included in Table 2.3.

In comparison with salamanders, species from most other taxonomic groups in freshwater or terrestrial habitats show much wider fluctuations in abundance over comparable periods. Species were selected on the basis of the duration of the study relative to the life history of the organism, rather than to the constancy of the population. They were a lamprey, a mayfly, a protozoan, five terrestrial insects, a lizard, two mice, and three birds. In general, the vertebrates displayed greater constancy of population size, as might be expected from their known ability to avoid or to tolerate environmental fluctuations. Even one of the birds was found to display rather large variations in abundance, however, thus showing that the group most involved in community theory is not sure to meet the assumptions of stability and equilibrium. Salamanders are shown to be among the organisms most suitable for testing community theory.

3

An introduction to the variety of salamander life histories

It is a common misconception, even among biologists, that nearly all salamanders are amphibious. Many people know about the neotenic axolotls (mostly Mexican members of the genus *Ambystoma*), but regard them as exceptions. The extent of the misconception is shown in Table 3.1. A considerable majority of all species of salamanders are completely terrestrial and undergo metamorphosis while still in the egg.

Within each of the categories in the table there is variation in the duration of each stage. The usual method of estimating the characteristics of life histories is to select some method of determining the ages of individuals collected in the field. Only in exceptional cases can this be done directly. The method used most frequently is to assume a relationship between size and age. Where the salamanders have distinct size classes at each time of collection, repeated collections can reveal growth rates, and a given size class can be followed for as long as it remains distinct. Typically, a collection contains three distinct size classes, representing the first year, the second year, and all subsequent years. Animals three years and older compose a single size class because the combination of initial size differences and different growth rates causes such an increase in the variance in size after the second year that year classes can no longer be distinguished by size alone. Sometimes an additional year class can be separated on the basis of sexual immaturity, but there is nearly always a large group of animals whose age must be estimated by very indirect methods. The exception is among the males of some genera of Plethodontidae, whose testes consist of a series of lobes, the number of which is assumed to represent the number of spermatogenic cycles experienced. Interpretation of the number of lobes is helpful, as compared with the use of size alone, but the process does have pitfalls, as described by Tilley (1977). For example, there is a limit to the number of lobes that can form along a spermatogenic cord, thus limiting the maximum age that can be estimated by this method.

In conjunction with the descriptions that follow, examples will be given of studies in which various methods were used to deduce the details of a particular life history.

Table 3.1. *Distributions of life histories among families of salamanders*

Family	Total species	Amphibious species	Species always neotenic	Some neotenic populations	Aquatic adults	Terrestrial species	Unknown
Cryptobranchidae	3				3		
Hynobiidae	32	22		3	3		4
Amphiumidae	3		3				
Proteidae	6		6				
Ambystomatidae	31	18	4	9			
Dicamptodontidae	3	1	1	1			
Salamandridae	53	21		6	17	2	7
Plethodontidae	221	19	6	4	2	189	1
Sirenidae	3		3				
Totals	355	81	23	23	25	191	12

Amphibious life histories

Ambystomatidae

In eastern North America, members of the genus *Ambystoma* spend the larval period in temporary ponds, metamorphosis coinciding approximately with the drying of the pond in summer. Typically, adults migrate to the ponds shortly after autumn or winter rains have filled them. There is, however, a sequence of arrival of the different species in most areas, and this sequence may have a profound effect on the nature of ecological interactions in the community, as described in Chapter 5. In the northeastern United States and eastern Canada, the sequence is, first, either *A. laterale* or *A. jeffersonianum*, which arrive before the ponds are completely ice-free, followed by *A. tigrinum* (where present), and then *A. maculatum*. South of the distribution of *A. laterale*, *A. opacum* lays eggs in early autumn, in depressions that will fill with water after leaf fall (late November). *A. opacum* has thus carried early breeding to an extreme. During winter and early spring, some of the same ponds are breeding sites for *A. talpoideum*, *A. tigrinum*, *A. maculatum*, and *A. mabeei* in various combinations.

The date of arrival at the breeding ponds is more variable for some populations than for others. In Michigan, Husting (1965) observed arrival seasons of *A. maculatum* as short as nine days and as long as twenty-nine days over five years. In North Carolina, Stenhouse (1984) intercepted the same species over

Table 3.2. *Monthly variations in migrations of* Ambystoma talpoideum *captured with a drift fence and pitfall traps enclosing a Carolina bay near Aiken, S.C., between 16 February 1975 and 31 May 1976*

	1975		1976	
	In	Out	In	Out
January			92	54
February	178	1238	24	372
March	32	452	19	230
April	63	814	82	1111
May	27	288	1	14
June	1	22		
July	3	6		
August	0	0		
September	0	1		
October	0	0		
November	306	16		
December	806	29		

Source: Patterson (1978).

twenty-seven days in 1980 and fifty-two days in 1981. Two males that arrived thirty days before the rest of the population accounted for the unusual length of the 1981 season. The figures do not include some very late stragglers, which were probably wandering rather than engaging in a breeding migration. For the autumn-breeding *A. opacum*, Stenhouse (1984) recorded a season of fourteen days in 1980 and twenty-six days in 1981. In contrast to those two species, *A. talpoideum* in South Carolina has a prolonged migratory season. Patterson (1978) observed appreciable immigration starting in November and running through May (Table 3.2). The duration of the season of migration is apparently influenced by local conditions at least as much as by characteristics of the species. Shoop (1960), working with *A. talpoideum* in Louisiana, observed a period of migration of only six to fifteen days over a four-year span.

There is an obvious advantage to early breeding. Eggs hatching with the first rain allow the larvae to grow to an advantageous size before those of the later-arriving species can develop. They not only have the first opportunity to prey on the small aquatic fauna but also are often large enough to be predators on smaller salamander larvae. The habit of *A. opacum* of laying eggs before the ponds fill with water does expose them to drought when it occurs and to any extreme cold weather that may occur before the rain arrives.

The number of adult *Ambystoma* migrating to breeding ponds varies much more widely from year to year than would be expected from their known longevity. Husting (1965) observed a range in numbers of male *A. maculatum* from 111 to 315 over five years in the same pond in Michigan. The range in numbers of females caught at the same fence was from 69 to 155. Maximum numbers of the two sexes did not appear in the same year. Husting estimated that in an average year, 36 percent of the males and 32 percent of the females migrated to the pond. His figures were based on a total population estimated by mark–release–recapture. Gill (1985) offers a different interpretation, based on his experience with the newt *Notophthalmus viridescens*. In contrast to Husting, Gill had complete fences and was able to estimate the proportion of animals that he failed to capture through their "trespassing" over or under the fences. He could account for nearly all of the failure of his animals to breed in some years, and he suggested that Husting's data could be understood in the same way. His interpretation requires significant year-to-year differences in the level of trespassing.

Shoop (1974) observed a wider range in numbers (16–67 females in different years) in a pond in Massachusetts. Because his data showed a regular decline, he attributed the differences to his fencing the pond and handling the specimens. The regularity was not observed by Husting, who caught the largest number in the last of five years. The same was observed by Whitford and Vinegar (1966) in Rhode Island. The irregularity of appearance in different years is undoubtedly related to the occurrence of rain during a relatively short time in late winter or early spring, when conditions are otherwise suitable for breeding. C. R. Shoop (pers. commun.) believes that the weather must be too cool for snake activity, as snakes are predators on migrating salamanders. This would set a limit to how late *A. maculatum* could breed, regardless of how long the pond in question retained water.

The most complete single study of any species of *Ambystoma* is that conducted on *A. talpoideum* at the University of Georgia's Savannah River Ecology Laboratory in South Carolina, mostly by Semlitsch (1981, 1983b, 1985a,b, 1987), but also by Patterson (1978) and by Semlitsch and Gibbons (1985). Like *A. maculatum*, *A. talpoideum* depends on sufficient rainfall during the critical period prior to actual breeding, and the number of animals caught at fences varies greatly between different years. Table 3.3 shows year-to-year fluctuations at three nearby ponds. Sun Bay was mostly drained in 1978, and hence the returning salamanders were adults before then, as the remaining

Table 3.3. *Breeding population and annual juvenile production at three breeding sites of the salamander* Ambystoma talpoideum *in South Carolina*

Site and year	Breeding adults M	F	Estimated egg production	Metamorphosing juveniles	Production (juveniles per female)	Survival to metamorphosis (%)
Rainbow Bay						
1978–9	243	225	98,529	4,029	17.9	4.09
1979–80	1,072	1,055	437,960	2,924	2.8	0.67
1980–1	247	86	33,019	3	0.03	0.01
1981–2	425	500	228,394	8,136	16.3	3.56
1982–3	1,248	827	312,331	4,624	5.6	1.48
1983–4	1,531	1,065	414,258	9,512	8.9	2.30
Sun Bay						
1978–9	469	531	262,725	0	0	0
1979–80	3,013	3,300	1,830,003	0	0	0
1980–1	113	53	31,695	0	0	0
1981–2	1,009	974	566,221	0	0	0
Bullfrog Pond						
1979–80	—	—	—	560	—	—
1980–1	95	28	12,656	0	0	0
1981–2	119	138	64,581	820	5.9	1.27

Source: Semlitsch (1987).

shallow pond dried each year before the larvae could metamorphose. The fluctuations there were almost certainly not related to recruitment from previous years, because this species reproduces at the age of one year, and there would have been no source of new adults.

Survival of the eggs in the field was not monitored, as they are scattered singly over the pond bottom, but one source of mortality was tested experimentally in the laboratory (Semlitsch, pers. commun.). He placed the eggs in aquaria with a common local species of fish, *Lepomis macrochirus*, and found that they were invariably consumed, in contrast to the eggs of *A. maculatum*, which are encased in an outer jelly envelope, which the fish were unable to open, despite repeated attempts.

Survival of the eggs of other *Ambystoma* species has been monitored. Stenhouse (1984) followed seventy-four egg masses of *A. maculatum* in North Carolina. She found much variation in viability from pond to pond, with an average survival to hatching of 48 percent. In some of the same ponds, *A.*

Table 3.4. *Summary of body sizes and reproductive traits of known-age female* Ambystoma talpoideum *from two breeding seasons (1981–3) in South Carolina*

Age (years)	*N*	Snout–vent length (mm)	Number of ova	Diameter of ova (mm)
1	45	46.8 (1.08)[a]	248 (21)	1.19 (0.05)
2	64	53.1 (0.57)	275 (14)	1.20 (0.01)
3	35	57.7 (0.49)	445 (27)	1.35 (0.02)
4	33	61.4 (0.41)	516 (27)	1.38 (0.01)
5	21	62.4 (0.45)	668 (36)	1.53 (0.01)

[a]Mean ± 1 SE.
Source: Semlitsch (1985a).

opacum, which deposits its eggs before the ponds fill with water, had an egg survival of 36 percent.

Semlitsch estimated the fecundity of *A. talpoideum* by collecting sexually mature females and counting the number of yolky follicles in animals entering the pond, and also in those leaving after the breeding season. There was a relationship between fecundity and age, with one-year-olds having an average of 248 ova, and those five years old and older having 668. The details are given in Table 3.4. Older females lay larger eggs, which probably gives an advantage in faster early growth and more certain attainment of a size large enough to be sure of metamorphosing when the pond dries up.

Survival from egg laying to metamorphosis varied between years from 0.01 to 4.09 percent. This was not related in any obvious way to the number of eggs that Semlitsch estimated had been laid, and he concluded that density-dependent effects were unimportant during larval existence. However, if we ignore the disastrous drought year 1980–1, when the pond contained water for only 116 days, we find that the negative regression of number of young produced per female on number of breeding females is statistically significant: $r^2 = 0.790$ ($F_{1,3} = 11.319$, $p = 0.044$). This finding suggests that larval survival is sensitive to density in most years. Other authors have reported similar results. Stenhouse, Hairston, and Cobey (1983) showed that growth of larval *A. opacum* was inversely related to experimental densities in field enclosures, when the densities were realistic in terms of known field densities, and there is a similar trend in the data of Shoop (1974) for *A. maculatum*. Semlitsch (1983a), however, did not find any evidence for density-dependent factors in his study of the life history of *A. tigrinum*.

Table 3.5. *Life table for* Ambystoma talpoideum *in Rainbow Bay, Savannah River Ecology Laboratory, South Carolina*

x (years)	l_x	m_x	$l_x m_x$
0	1.0000		
0.5	0.0202		
1.0	0.0028	125	0.35
2.0	0.0016	137.5	0.22
3.0	0.0009	222.5	0.20
4.0	0.0005	258	0.129
5.0	0.0003	334	0.10
6.0	0.0001	334	0.033
			$R_0 = 1.032$

Source: Semlitsch (1985a, 1987, unpublished data).

The survival of adults can only be inferred by mark–release–recapture of migrating individuals. For *A. talpoideum*, Dr. Semlitsch has kindly provided unpublished estimates from his study of the population in Rainbow Bay. As described earlier, not all metamorphosed individuals return to breed every year. The estimates derived from recaptures are therefore of minimum survival. From metamorphosis to first breeding, the average survival over four years was 0.139, with a range from 0.035 to 0.192. Survival of animals that had bred at least once was much higher, and averaged 0.566 per year, with a range from 0.482 to 0.623. The data from the partly drained Sun Bay (Table 3.3) do not support estimates of a higher death rate, as there is little, if any, evidence of a decline in the adult population that was faithful to the pond, despite complete lack of success in producing young.

The data from Rainbow Bay are sufficiently complete to allow the construction of a life table for *A. talpoideum* (Table 3.5). The calculated net reproductive rate (1.032) reveals that the population was more or less stationary, with the principal deviations being due to irregular breeding and the resulting dominance by some year classes, 1979 and 1982 in this study. The data for the different parts of the life table were collected independently, a feature that gives confidence in the estimates of age-specific survival and reproduction.

Estimates of the survival of *A. maculatum* are higher than those for other *Ambystoma* species. Both Husting (1965) and Whitford and Vinegar (1966) obtained survival rates in the range of 80–94 percent per year. These figures imply very low annual production of young, because the populations did not appear to be increasing. The data of Stenhouse (1984) are consistent with that conclusion, as she observed only four metamorphs in two years from a population estimated to contain 128 females. Not all females migrated in both years, but Stenhouse estimated that 3,672 eggs were present in one year.

From the information presented, it can be seen that there is much variety in the details of life history among the species in a single genus of salamanders, all of which are classically amphibious.

Salamandridae

Like *Ambystoma*, most species of Salamandridae have aquatic larvae and terrestrial adults. In life history, European newts of the genera *Triturus*, *Pleurodeles*, *Euproctus*, *Chioglossa*, and *Salamandrina* differ from each other largely in the locations in which the eggs are laid. The first two breed in ponds or other still water; the remaining three are primarily stream breeders, although *Euproctus* may deposit eggs in shallow water in mountain lakes. The duration of the larval period depends, in several species, on when in the breeding season the eggs are deposited. This may cover several months in species of *Triturus*, and larvae hatching late may overwinter before metamorphosis. Otherwise, the larval stage lasts for three or four months. *Pleurodeles* is unique in having no definite breeding season. The remaining European salamandrids, *Salamandra salamandra* and *S. atra*, are ovoviviparous. *S. salamandra* deposits active larvae in streams or stream-fed pools; in *S. atra*, metamorphosis takes place within the oviduct, and the terrestrial young are born with the adult form.

A thorough study of the life history of the smooth newt, *Triturus vulgaris*, has been carried out by Bell (1974, 1977) and Bell and Lawton (1975). In the Oxfordshire ponds where Bell worked, there were frequently three more-or-less distinct periods of egg laying each year. Figure 3.1 shows the distribution of oviposition during one season in one pond, with the "main" cohort clearly dominating. The early period was centered near 23 March. That cohort contained 15 percent of the total. The main cohort, which represented 77.5 percent of the eggs, peaked about 7 May; the late cohort, peaking at 2 June, contained only about 7.5 percent of the eggs. This pattern was not followed in all four ponds that were observed. In two of them, only single size classes of larvae were seen.

Two different methods were used in successive years to estimate the survival of eggs from oviposition to hatching. Survival improved up to the middle of the laying of the main cohort, an observation interpreted by Bell and Lawton to demonstrate the adaptive nature of the timing of most of the

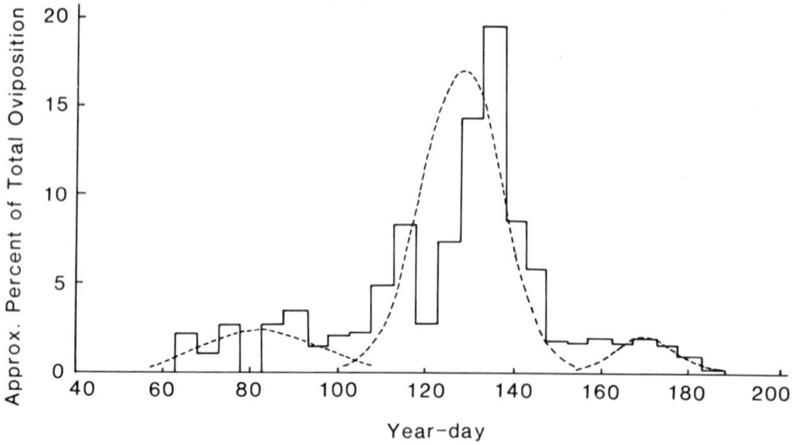

Figure 3.1. Time course of oviposition at N.C. Arboretum in 1972. Symmetrical normal curves have been fitted to the data by the method of Harding (1949). (From Bell & Lawton 1975.)

oviposition. The two methods (direct sampling of the pond to follow changes in the age structure of a total of 1,218 eggs, and following 277 laboratory-laid eggs in the field) gave excellent agreement in estimating the survival of eggs as 0.0263 and 0.0255, respectively.

At least two size classes of larvae were discernible in both 1971 and 1972 in the pond from which the egg-laying data were taken (Fig. 3.2), but the late cohort in 1971 was much more prominent than it was in the following year, which is the one ilustrated in Figure 3.1. Survival of larvae was estimated from the size-frequency data taken at successive weekly intervals. The result of the calculations was that survival from hatching to metamorphosis was 0.0905. This figure agreed well with direct counts of samples taken early and late in one season in a single pond, where 10 percent were found to survive.

Analysis of the terrestrial parts of the life history depended largely on trapping migrants to and from the ponds and examining the specimens, either by anesthetizing them and measuring them prior to release or by preserving them and dissecting for gonad examination. The former approach yielded larger amounts of data; the latter gave precise information on reproductive condition.

The use of size-frequency data to estimate age classes of adults migrating to the ponds to breed was facilitated by calculating five-point moving averages of the data in Figure 3.3. Pond 1 in Figure 3.3 yielded the successive peaks in Figure 3.4, which show much more regularity, and which Bell interpreted as

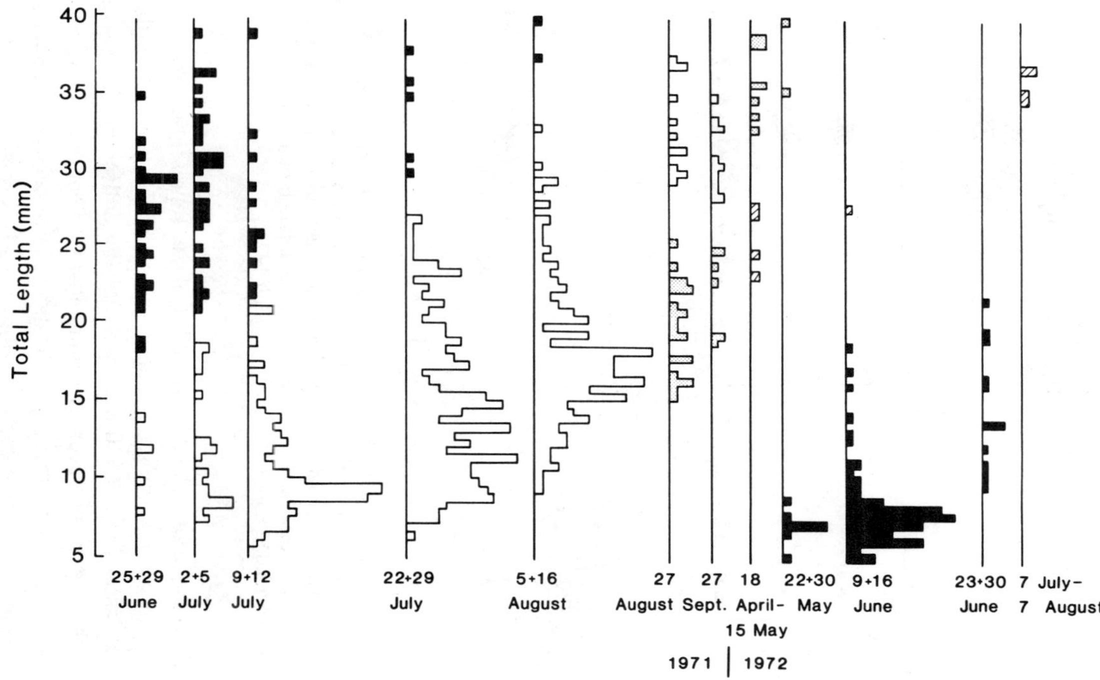

Figure 3.2. Length distributions of larvae in N.C. Arboretum, 1971–2 (N = 837). Hatched squares, early cohort; solid squares, main cohort; open squares, late cohort; stippled squares, overwintering group of late-cohort animals. (Adapted from Bell & Lawton 1975.)

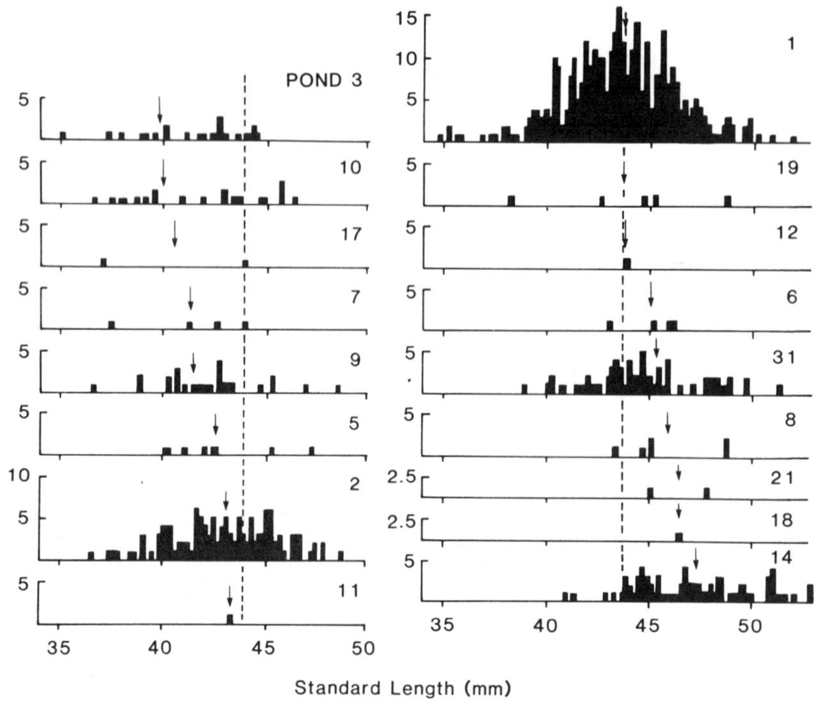

Figure 3.3. Frequency distributions of samples of adult male smooth newts from main-study-area localities taken in 1972. Raw data lumped at 0.2-mm intervals of standard length. An arrow indicates the mean for each sample; weighted mean of means shown by broken line through middle of diagram; abscissas divided at intervals of 5. (From Bell 1977.)

representing successive age classes, including those representing the two principal cohorts of metamorphosed young within years. He was thus of the opinion that age classes 4 through 10 were present in the collection. There was one problem with the data, however. Assuming that no great fluctuations had occurred in survival and reproduction, there should have been more animals in each younger age class than in any succeeding one. This was clearly not the case, and his interpretation was that there was a large span of years when some newts were immature and others mature within the same age classes. Figure 3.5 shows the correctness of the hypothesis. In several of the collections, animals of the same size class fall into two groups: those that are still immature, and those that are ready to breed. Bell was able to calculate the proportion becoming mature in each successive year from age three to age seven, as shown in Table 3.6.

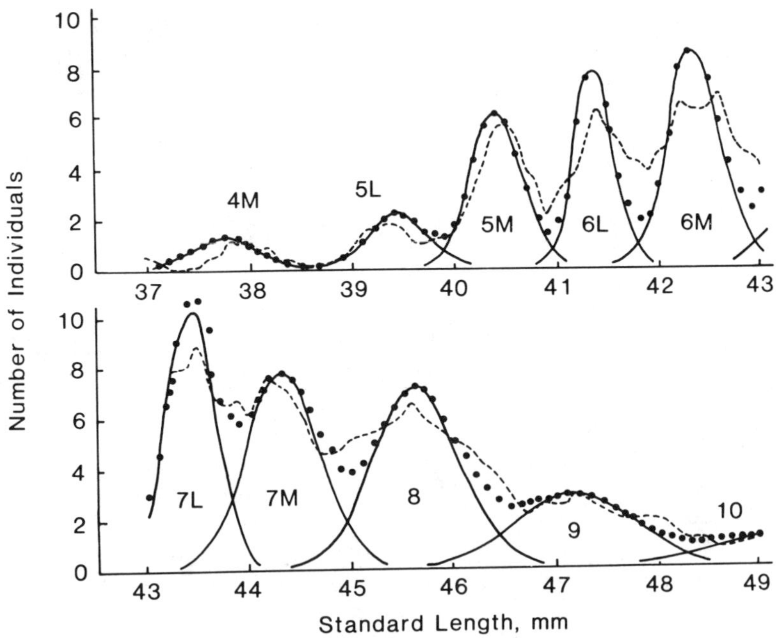

Figure 3.4. Dissection of the N.C. Arboretum 1972 male length-frequency distribution into its supposed constituent distributions. The broken line is a simple five-point moving average of the raw data; the unbroken lines are the supposed constituent distributions, whose sum is represented by the points. M and L refer to the age classes contributed by the main and late oviposition periods, respectively. (From Bell 1977.)

Bell estimated that the annual survival of immature terrestrial animals was about 80 percent (Bell & Lawton 1975), and several estimates of adult survival have an average of 50 percent per year (Bell 1977). This fits well with Bell's finding that very few adults live longer than ten years. Adult females begin by laying about 100 eggs per season. Large individuals may lay as many as 400.

The data are now complete enough to produce a life table. The result indicates that the population should have been declining, if all of the estimates of age-specific survival and fecundity are correct. This is shown in the calculation of the parameter R_0, the net reproductive rate per generation, which he calculated to be approximately 0.33. This seems low, but it is not obvious where any error occurred of sufficient magnitude to account for the difference between the calculated value and the value of 1.0 required for a stationary population. Intuitively, the loss of 97.5 percent of eggs before hatching ap-

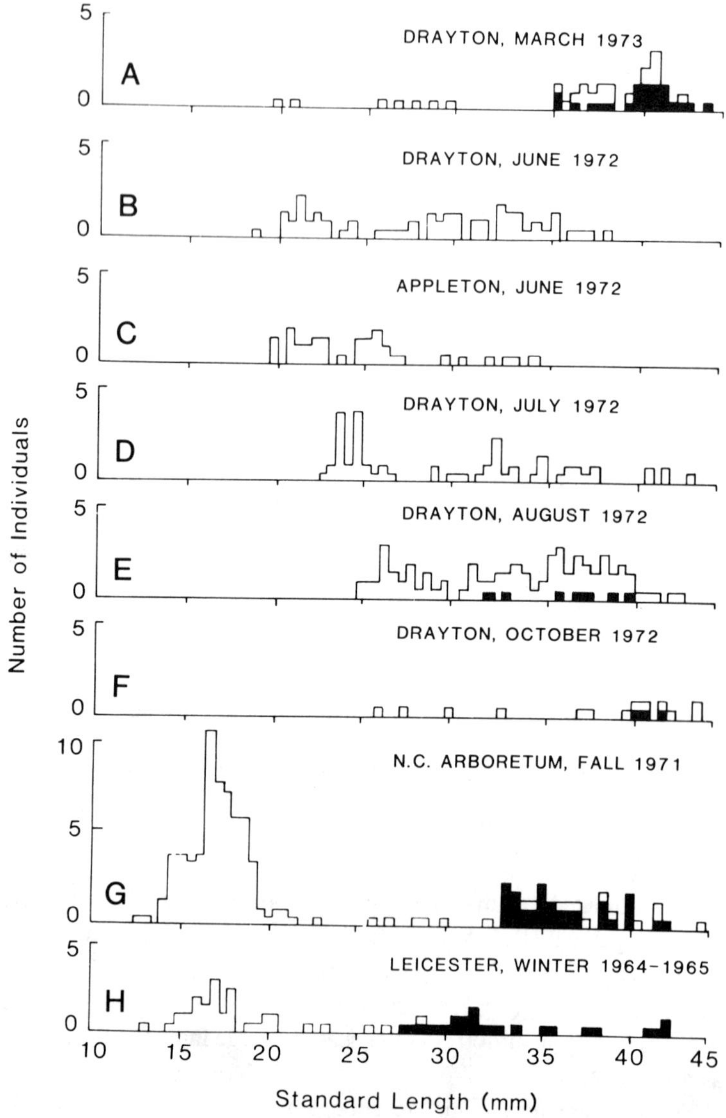

Figure 3.5. Standard length-frequency distributions of samples of terrestrial newts: (A) migrating adults and some juveniles; (B) juveniles only; (C) juveniles only; (D) juveniles only; (E) juveniles and some newly mature adults; (F) juveniles and migrating adults; (G) metamorphosing juveniles, juveniles, and migrating adults; (H) juveniles and breeding adults. Solid squares represent individuals shown by dissection to be sexually mature. (From Bell 1977.)

Table 3.6. *Schedules of maturity in two populations; the variances of the estimates are unknown*

	Proportion of animals first reproducing at age *a*			
	NC Arboretum		Drayton	
Age *a* (years)	M	F	M	F
2	0	0	0	0
3	0+	0.0066	0.1419	0.1372
4	0.0740	0.0163	0.7112	0.4048
5	0.2296	0.2070	0.1159	0.4581
6	0.3795	0.2244	0.0311	0+
7	0.3169	0.5457	0+	0+
8	0+	0+	0+	0+

Source: Bell (1977).

pears high, and Bell commented on unusual predation by caddis flies on eggs and small larvae in the year when the best data were obtained. Bell's estimates of the sizes of the populations, based on recaptures, do in fact show a marked drop in numbers between the second and third of the four seasons of his study, and the expectation implied in the calculation of the net reproductive rate may be justified.

Further information on the life histories of other European salamanders can be found in Steward (1969) and Frazer (1983).

Variations in the life histories of European newts are less dramatic than, but similar to, those found in the life history of the newt of eastern North America, *Notophthalmus viridescens*. This species is usually thought of as having aquatic adults, a larval period of several months, and a terrestrial immature but metamorphosed stage (the eft) lasting three to six years or more. A common variation is for the adults to undergo an annual migration away from the breeding ponds in late summer, with a return in early spring (Gill 1978). The pattern makes these populations similar to the European *Triturus*. In a number of areas on the coastal plain, the eft stage is omitted, presumably because the surrounding sandy habitat is inhospitable to terrestrial specimens. In several coastal-plain ponds, the populations have been observed to consist largely of neotenic individuals, with a few normal adults. All newts are voracious predators, and their impact on pond communities must vary with the type of life history exhibited, ponds from which the adults migrate annually being less affected than those that are inhabited continuously. Some of this impact is described in Chapter 5.

Plethodontidae

Among the amphibious Plethodontidae, all of which are confined to eastern North America, there is great diversity in the relative time spent in the larval stage. This has been found to be as short as a few weeks, but is more commonly at least several months. *Eurycea longicauda* lasts three to six months as a larva (Bruce 1982), whereas *E. bislineata* is a larva for one year in some parts of the southern Appalachians, and two years in others. Bruce concluded that the shorter period is related to the large number of competing species with which those populations coexist. That would be consistent with Wilder's observation (1924) that, in Massachusetts, the larval period is two or three years because the only potential competitor there is *Desmognathus fuscus*.

The longest larval period of any plethodontid is that of *Gyrinophilus porphyriticus*, which has been found to metamorphose at a modal age of four years, with a range in larval life of from three to five years (Bruce 1980). For terrestrial salamanders in the vicinity of the streams inhabited by *Gyrinophilus*, the long larval life is a blessing, as the adults are specialists in feeding on other salamanders.

Permanent larvae

As seen in Table 3.1, five families of salamanders have at least some species that are neotenic. Moreover, the Cryptobranchidae and the Amphiumidae do not metamorphose completely. Reversion from a presumed ancestral amphibious life to an entirely aquatic life has been widespread. The correlates of neotenic life are as follows: inhabiting subterranean waters, inhabiting ponds or lakes lacking in fish predators, inhabiting lakes or large streams, and being surrounded by hostile terrestrial habitat. When obligate, the first correlate is perfect. The others are only suggestive, and none applies to the Amphiumidae and Sirenidae, all members of which inhabit ditches, swamps, and sluggish streams in the southeastern United States. This area is the home of at least a dozen species of salamanders that metamorphose normally, and here one must also look to some factor other than habitat for an understanding of the neotenic condition (Patterson 1978).

Many species of Ambystomatidae and Salamandridae are facultatively paedomorphic, with some individuals reproducing while retaining the larval morphology. The proportion may vary within a population from year to year, depending on conditions in the aquatic environment (Healy 1970; Semlitsch 1987).

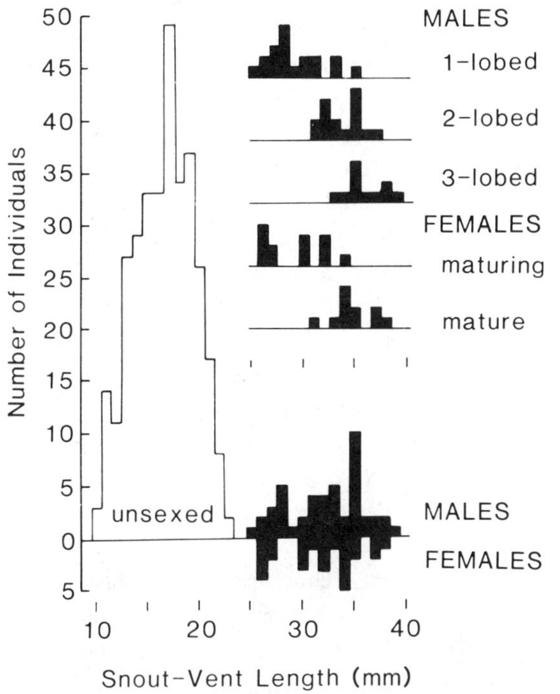

Figure 3.6. Distribution of snout–vent lengths of 393 *Eurycea neotenes* taken between 25 July and 8 August 1970. Open squares represent immature, first-year individuals; solid squares represent older individuals. (From Bruce 1976.)

Unfortunately, there appear to be no studies of neotenic forms that are as complete as those by Semlitsch and Bell, but Bruce (1976) has published data showing the population structure and age-specific survival of *Eurycea neotenes*, a plethodontid in central Texas. This species inhabits springs and the outflow streams arising at the base of the Edwards Plateau, a limestone formation lying over a more impermeable base. The surrounding terrestrial habitat is dry (70–80 cm annual rainfall) and unsuitable for these salamanders, the relatives of which metamorphose normally where they occur in wetter climates. Indeed, *E. neotenes* will metamorphose when treated with thyroxine (Kezer 1952).

Two populations were sampled in late July and early August. One collection consisted of 393 specimens, the other of 215. The size structure of the larger collection is shown in Figure 3.6. The figure includes information bearing on the age structure of the larger animals. Males of this species have lobed testes, and Bruce concluded that each lobe represents a season of reproductive life, except for the three-lobed group, which includes all individuals in their third year of reproduction and later. Those in the large group of small unsexed animals were in their first year, having hatched from eggs that were laid during the previous winter. Males with one lobe per testis belonged to the next larger size class, as did the group of maturing females. Bruce interpreted his data to mean that males are sexually mature one year after being laid as eggs, and females lay their first eggs at two years of age. The life history, as far as age at maturity is concerned, is much like that of the metamorphosing relative *Eurycea bislineata* in the southern Appalachians (Bruce 1982).

Survival was poor in the first year of life; Bruce (1976) estimated it as 0.099 in one population and 0.083 in the other. After maturity, survival was higher, and though an exact calculation is not possible, a crude estimate of 0.6 per year is consistent with the data. No observations were made of fecundity for *E. neotenes*, but the large number of first-year specimens per surviving female suggests that it is high.

Metamorphosis appears to entail considerable risk (Just, Kraus-Just, & Check 1981), both from the extensive reorganization of tissues (De Jongh 1968) and from the temporary vulnerability of metamorphosing animals to predation (Arnold & Wassersug 1978). Thus, there is an advantage in survivorship in cases where the larval form is retained into sexual maturity.

One might expect that the omission of metamorphosis would shorten the time required to reach sexual maturity. There would be an obvious advantage, from the standpoint of increasing the intrinsic rate of natural increase, in shortening the mean generation time (Cole 1954; Lewontin 1965). When we compare the life histories of neotenic and metamorphosing forms of the same species, however, we find that the time required to reach maturity is the same. For example, Glass (1951) found that neotenic *Ambystoma tigrinum* was sexually mature at one year of age, the same age required by metamorphosing individuals (Bishop 1941). The same is apparently true of *Dicamptodon ensatus*, a large ambystomatoid of western North America (Nussbaum 1976). Gould (1977), in agreement with Wilbur and Collins (1973), interpreted this failure of permanent larval forms to shorten the generation time as indicating that these species are under density-dependent ecological conditions that would favor characters other than those increasing the intrinsic rate of natural increase. This implies that "they inhabit highly favorable ponds with stable, non-expanding populations" (Gould 1977, p. 320). As shown in Chapter 2, such populations are typical of salamanders generally, whether neotenic or not. In contrast to that conventional interpretation, Sexton and Bizer (1978) pointed to their finding of a correlation between low temperatures and increasing frequency of neoteny in *Ambystoma tigrinum* in the Rocky Mountains of Colorado. Snyder (1956) had reported earlier a similar trend in *A. gracile* in the Cascade Mountains of Washington State.

Confirmatory evidence for the conventional interpretation comes from the long immature life of the large neotenic salamanders of the genus *Necturus*. *N. maculosus* becomes adult at six years (Bishop 1941). The same is true of *N. lewisi*. The smaller *N. punctatus* matures in five years (Martof et al. 1980).

One species of neotenic salamander, *Proteus anguinus*, which lives in caves and other subterranean water in Yugoslavia, is ovoviviparous. The immature period is not known. The taxonomic placement of *Proteus* in the same family as the American *Necturus* has been the subject of some debate. In Table 3.1, the two genera are treated as confamilial, without any serious implication that I endorse that side of the argument. The issues are well presented by Larsen and Guthrie (1974) and by Hecht and Edwards (1976).

Aquatic salamanders with metamorphosis

Two species of plethodontid salamanders have life histories that appear to retain the disadvantageous features of metamorphosis without the benefits of an optional terrestrial stage. They are *Stereochilus marginatus* and *Leurognathus marmoratus*. Both are completely aquatic as adults. It is not at all clear why they have not become neotenic. Their habitats are not alike. *Leurognathus* is confined to swift mountain streams in the southern Appalachians; *Stereochilus* is found in drainage ditches, small ponds, and sluggish streams on the coastal plain of the southeastern United States. Other salamanders have some populations that are similar to these two species, but there are obvious ecological correlates with the habit. *Notophthalmus viridescens* has been investigated in Massachusetts by Healy (1970, 1974). Ponds on Cape Cod and elsewhere on the coastal plain contain populations of newts that have no terrestrial stage. At least some of these populations contain neotenic individuals, the proportion varying from year to year, apparently depending on conditions in the ponds. Healy attributed the omission of the terrestrial stage to the harsh and unpredictable terrestrial environment, where the sandy soil provided an unsuitable habitat for terrestrial salamanders. Populations from inland locations were typical of the species in their life history, with efts remaining terrestrial for up to six years. Thus, the adoption of a completely aquatic life history is an intermediate step toward neoteny, and is unlike the life histories of *Leurognathus* and *Stereochilus*.

Bruce (1971) has described the life history of *Stereochilus*. Using size-

Table 3.7. *Seasonal variation in the size at which metamorphosis occurs in* Leurognathus marmoratus, *as shown by snout–vent measurements (mm); all specimens were collected in Jaccard's Creek, Lumpkin County, Georgia*

	Larvae			Subadults (40 mm)		
Date (1954)	Number collected	Size of smallest	Size of largest	Number collected	Size of smallest	Average size at metamorphosis
May 8	11	15	37	2	36	36
May 30	24	14	36	12	30	34
June 6	17	14	32	5	33	32
June 13	16	16	32	12	33	32
July 10	14	20	29	6	31	30
Aug. 7	15	21	30	17	30	30
Sept. 11	39	12	31	49	28	30
Oct. 10	21	12	31	6	29	30

Source: Martof (1962).

frequency data, he concluded that larvae, hatching in early spring, metamorphose at an age of twenty-five to twenty-eight months. Sexual maturity was coincident with metamorphosis in males, which were in reproductive condition at thirty-three months of age. Most females remained immature for a year after metamorphosis and deposited their first eggs when four years old, after which they reproduced annually, unlike some species of *Desmognathus* and *Plethodon*, which are biennial breeders, as described later. The long larval period is more like that of other plethodontids than it is like that of other pond-breeding salamanders. The American salamandrid *Notophthalmus viridescens* undergoes metamorphosis at the age of six months. In populations that remain aquatic, sexual maturity is attained at two years, half the age of female *Stereochilus*. In contrast, in newt populations that metamorphose into the terrestrial eft stage, maturity is delayed for a number of years, up to six, at least (Healy 1974; Gill 1978).

The other completely aquatic but metamorphosing plethodontid, *Leurognathus marmoratus*, has been studied by Martof (1962) and by Bruce (1985). Because it belongs to a different subfamily, its acquisition of a purely aquatic life was evidently independent. Eggs are laid in May and June; they hatch in September and October, the newly hatched larvae being about 11 mm in snout–vent length. Martof's data on the range of sizes in different months of the active season are given in Table 3.7. He interpreted the figures to mean that metamorphosis occurred in August after as little as ten months of larval life, with some late-hatching individuals overwintering a second time to metamorphose in May, after twenty months. The data in his table do not sustain that interpretation. Just before the hatching season in early August, the smallest larva was 21 mm in snout–vent length. Evidently, it had attained that length nearly one year after hatching. But metamorphosis takes place at a snout–vent length of 30 mm, the size of the largest larva and of the smallest newly transformed specimens in August. That leaves 9 mm of growth unaccounted for. It seems probable that another year of growth is involved, the amount being comparable to growth during the period from hatching in September to August. This would mean a minimum of twenty-three months as a larva. Martof recognized that some metamorphosis occurs in May, the largest larva at that time being 36 mm in snout–vent length, the same length as the smallest transformed individual. Such specimens have spent another winter as larvae, as he recognized, but by the foregoing interpretation their age would be thirty-six months since being laid as eggs, or after thirty-three to thirty-four months as larvae. Bruce (1985) has recently made more extensive collections of *Leurognathus* over a longer time than did Martof, and the foregoing interpretation is in agreement with his.

Martof did not comment on the duration of the period between metamorphosis and sexual maturity. The smallest mature specimens in his largest collections were more than 50 mm in snout–vent length. To reach that size from 36 mm would require two years, and from 30 mm an additional nine months. The age of first reproduction is therefore five years from being laid as an egg, and it is possible that females require an additional year, as no specimen smaller than 55 mm contained yolky eggs.

Clutch sizes averaged 38.1 for twenty-one clutches found in the field. Martof commented that that number is close to the number of eggs in gravid females and concluded that survival of eggs is very good. As in all other plethodontids, the female remains with the eggs, presumably guarding them against predation. No information is available on the survival during the rest of the life cycle.

Terrestrial salamanders

With the exception of two salamandrids, including the ovoviviparous *Salamandra atra*, which was described earlier, all species of salamanders with no aquatic stage belong to the Family Plethodontidae. This is an anomalous situation, inasmuch as the plethodontids are lungless, which would lead one to expect that terrestrial life might cause respiratory difficulties.

Lunglessness is associated with the highly oxygenated mountain-brook habitat, where lungs would be disadvantageous in that they act as floats and would cause their possessors to be swept away into a dangerous rocky torrent. Thus, all lungless salamanders are inhabitants of such habitats, as are some of the European salamandrids, or are thought to be descended from inhabitants of mountain streams, as is the case with the Plethodontidae (Dunn 1926).

The origin of terrestrialism in amphibians implies that the terrestrial habitat has been favorable during the origin and persistence of the terrestrial species. Favorability in this case means generally high humidity, access to underground retreats, and an acceptable range of temperatures. The origin of the Plethodontidae can be placed geographically in eastern North America, from the fact that all of the species in the family with aquatic larvae are confined to that area (see Chapter 7). The ecological requirements indicate that Appalachia was the place of origin, because the area has been elevated since the close of the Paleozoic Era, thus providing both the mountain streams and the moisture-trapping topography that would permit the evolution of terrestrial lungless salamanders. The fact that terrestrialism has arisen independently at least three times (in the Tribes Plethodontini and Bolitoglossini and in the Subfamily Desmognathinae) demonstrates the requirement of a long history in the area of origin. This is confirmed by the distances over which some of the terrestrial forms have spread. The Plethodontini are found discontinuously from eastern to western North America, with two genera (*Plethodon* and *Aneides*) having species in both areas. The Bolitoglossini have the most remarkable distribution, with those in western North America, Mexico, Central America, and large parts of South America easily visualized as being the results of continuous distributional processes. The surprising example is the genus *Hydromantes*, with two species in Italy and Sardinia and three in California.

As an example of a detailed life history of a terrestrial salamander, my study of *Plethodon jordani* will be used (Hairston 1983b). The data were accumulated over eight years. Large samples were collected, and measurements were made on live animals by placing them on a glass plate under a towel, alongside a clear plastic ruler, and photographing them from below. When the photographs were printed, the image of the ruler was cut out and used to measure the length of head plus body (always referred to as snout–vent length). The sequentially collected size distributions were used to estimate growth and, in conjunction with examination of gonads of dissected specimens, were used to determine age at maturity. Length-frequency histograms for the first two years are shown in Figure 3.7. For most collections, three size classes can be identified. The smallest individuals were collected in

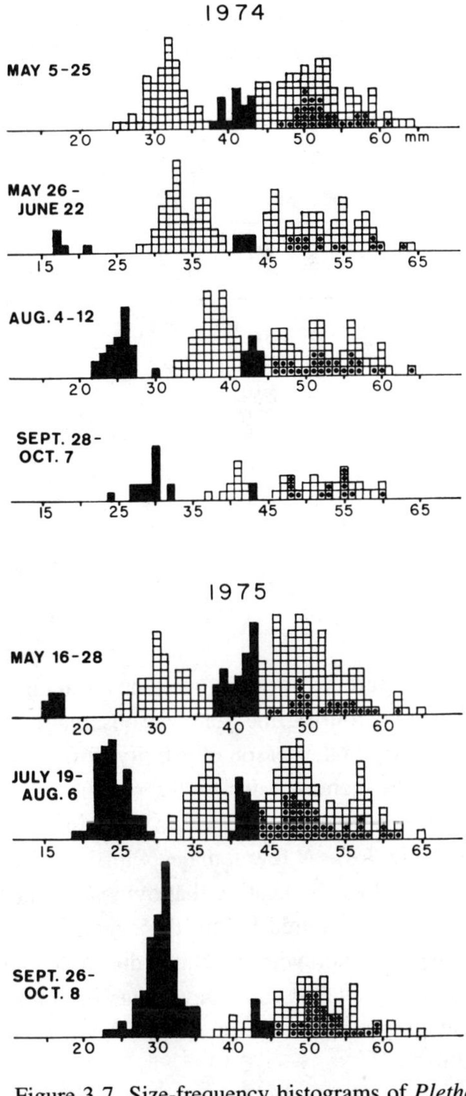

Figure 3.7. Size-frequency histograms of *Plethodon jordani* from the same locality during two active seasons. Closed symbols represent one-year-olds and that part of the largest size class estimated to be three years old. Open symbols represent two-year-olds and all specimens at least as large as the smallest externally identifiable male. Dots within open symbols represent such males. (From Hairston 1983b.)

Table 3.8. *Growth of eggs of* Plethodon jordani *during the late summer and spring; number of specimens with eggs in each category form the body of the table*

	Condition of eggs								
	No yolk			Some yolk			Dense yolk		
Egg diameter (mm)	Aug. 11	Sept. 24	May 6	Aug. 11	Sept. 24	May 6	Aug. 11	Sept. 24	May 6
4.00–4.59									1
3.50–3.99									
3.00–3.49								1	
2.50–2.99								4	
2.00–2.49							1	3	
1.50–1.99							8	4	
1.00–1.49	4			7	2	1	9	1	
0.50–0.99	7	15	10	3	2	9	1		
0.00–0.49	3		5						

Source: Hairston (1983b).

May of each year, and as the figure shows, they were never abundant. That size class increased in abundance during the year, an observation that was interpreted as indicating that early in the season of activity, few appear above ground, and that the proportion increases during the summer. These salamanders are one year old, as can be seen from the changes in the condition and size of ovarian eggs (Table 3.8). A few females with large yolky eggs were found in May, but none in June, indicating that oviposition takes place in May. Two to three months are required before the eggs hatch, and it is assumed that *P. jordani* females remain with the clutch during incubation, as do all other plethodontids (the eggs of *P. jordani* have never been found in the field). Thus, the small young animals that appear in May were hatched no earlier than the previous July.

The one-year-olds last seen in September–October are almost exactly the same size as the second size class is in May, when they are then two years old. Externally visible characters identify mature males. As Figure 3.7 shows, the smallest identifiable males are appreciably larger than the largest two-year-olds, especially early in the season. This permits identification of three-year-old animals, at least in May and June, even though they no longer form a separate size class. The number of adult age groups that are merged in the abundant group of large animals was estimated by making the assumptions that after the year in which males reach sexual maturity, survival is constant, and that the population was stationary. The first assumption is conservative, as the most likely deviation from a constant survival rate among adults would be for it to improve with age. That would increase the expected age of the older individuals in the size class, and their estimated age is already surprising. Adult survival was calculated to be 0.81 per year. As will be shown later, 11.9 percent of eggs survive to adulthood. The combined estimates mean that among a population of 1,000 animals, including eggs, one or two can be expected to be twenty-five years old or older. The assumption of a stationary population is shown to be approximately valid from counts on measured plots over a period of eight years (see Fig. 2.3).

Except for the first year of life, when the eggs are guarded by the female and when the newly hatched young stay below the surface of the ground until at least a year after egg laying, survival improves with age. By comparing the numbers in successive age classes at the same time of year, and knowing that the population was stationary, survival was calculated to be 0.364 from one to two years of age and 0.484 in the third year of life. Survival during the first year was estimated by constructing a provisional life table assuming perfect survival during that time. This gave a provisional net reproductive rate of 1.195. Inasmuch as the population data showed no trend in abundance over eight years, the rate should be 1.00. Losses during the first year being the only source of a correction, survival during that year was estimated as 0.837, the number that multiplied by 1.195 gives the required product of 1.00.

The fecundity part of the life table was determined from dissection of females, from which could be found the proportion gravid at any age and the number of yolky eggs in animals of different sizes, which were equated with age from a growth curve. No female lays eggs before the age of four, and during that and the subsequent year, only one-fourth of the females are reproductive. In the older age groups, half of the females are reproductive in each year. That situation comes about from the time required in guarding the eggs. Females cannot feed during the two or more months of incubation, and because hatching precedes the mating season by only a short time, those females are unable to make yolky eggs in time to participate and thus are unable to oviposit the following May.

In regard to variations in life histories, terrestrial plethodontids can be classified by whether they are viviparous or ovoviviparous, by the duration of the immature period, or by the seasonality of reproduction. *Hydromantes genei* of Sardinia bears living young, and there are claims (Noble 1931) that some of the tropical bolitoglossines are also ovoviviparous; all other species lay eggs.

The ages at sexual maturity for a number of terrestrial species are given by Houck (1977). The range for temperate-climate species is twenty-one months to a little more than four years for males; the maximum is ten months longer for females. These differences mostly reflect size differences, with larger species taking longer to mature. Among tropical species, the range is greater: two to six years for males and thirty months to twelve years for females. The very long immature period is for *Bolitoglossa subpalmata* from high elevations in Costa Rica (Vial 1968) and is documented by seventy-five marked individuals recaptured at least six months after marking. Presumably, the phenomenon is related to the unusually dense populations and possibly to the continually low temperature.

A genus with varied life histories: *Desmognathus*

A considerable proportion of the diversity in salamander life histories is found within the single genus *Desmognathus*. Previous authors had established some of the basic facts about the life histories of some of the species, but it was Organ (1961) who produced the first quantitative data on the five species with which he worked. He made massive collections; none of the species was represented by fewer than 1,000 individuals. The five species were *D. quadramaculatus*, *D. monticola*, *D. fuscus*, *D. ochrophaeus*, and *D. wrighti*. Organ found the eggs of all five species and estimated that they were deposited in early summer. His calculations of age were based on the assumption that all oviposition took place in June. Although not observed for all species, the hatching that was seen was in late summer or autumn. In those months for which the collections were large enough, *D. quadramaculatus* had larvae in two size classes. From June to September, the larger larvae overlapped in size with the newly metamorphosed salamanders that were appearing in the collections in late summer and autumn. He concluded that the larval period was two years. Three of the remaining species had only one size class of larvae. *D. monticola* metamorphosed in June and July, indicating that larval life lasted for eleven or twelve months. *D. fuscus* had a variable larval period, lasting between eleven and fourteen months. *D. ochrophaeus* metamorphosed in April or earlier, eight to nine months after hatching. *Desmognathus wrighti* had no larval stage. Metamorphosis took place before hatching. The duration of the larval period was correlated with the habitat of the adults. *D. quadramaculatus* is the most aquatic of the species. *D. monticola* and *D. fuscus* are streambank species, spending part of the time in the water and part in the forest, but are normally found in the noninundated part of the streambed. *D. ochrophaeus* is commonest in the forest close to the streambank, which it visits to some extent, and *D. wrighti* is completely terrestrial, except during the winter, when it remains in seepage areas.

Organ obtained estimates of the ages of the adult males of all of the species by counting the number of lobes on the testes. He assumed that males with a single lobe were either three or four years old, those with two lobes either five or six, and so forth. These assignments were based on the prior observation (Humphrey 1922) that two years are required to add one complete lobe. The numbers of animals in the several categories thus automatically give survival rates under the assumption of stationary populations. As noted earlier, the interpretation of numbers of lobes is more complicated than Organ supposed, but his basic conclusions remain.

The most interesting and important aspect of the estimates of survival of the different species is the relationship between the shape of the survivorship curves and the sizes of the species and their locations on the aquatic–terrestrial gradient. In representing these curves, it is conventional, following Deevey (1947), to plot the number or proportion remaining alive on a logarithmic scale on the vertical axis and age on the horizontal axis. A straight line denotes survival of a constant proportion during each time period. Curves that become increasingly steep with age are indicative of good survival early in life, with a change to a considerably higher death rate that is frequently associated with a more-or-less fixed life span. Figure 3.8 shows Organ's results for the five species of *Desmognathus*. *D. quadramaculatus* and *D. monticola* come close to straight-line relationships, with *D. monticola* having a slightly better average survival. The other three species show the second type of survivorship curve to a degree that increases from *D. fuscus* to *D. ochrophaeus* to *D. wrighti*. Organ used the name *carolinensis*, which was the then-current name for the southern subspecies of *D. ochrophaeus*. The curves apply only to postmetamorphic individuals and show that survivorship improves with an increasingly terrestrial existence. This relationship is shown to have interesting implications for interspecific relationships within the *Desmognathus* guild (see Chapter 6).

Females of each of the species were equally abundant with males up to the onset of maturity, as immature females and males with one lobe on the testis did not differ in numbers. The situation was different among mature females of the three smaller and more terrestrial species. This led Organ to conclude that mature females had a higher death rate than mature males of those three species. The females of all *Desmognathus* species lay their eggs in or near aquatic sites, and reasoning from the differential survival of males according to habitat, Organ concluded that the deposition and guarding of the eggs involved extra risk. Like *Plethodon*, the female *Desmognathus* in his study

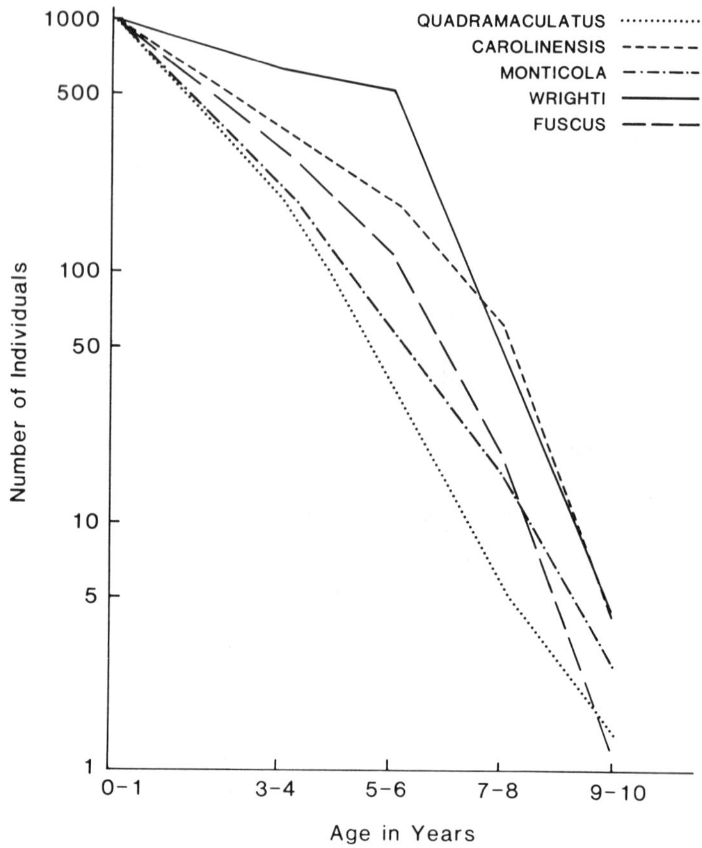

Figure 3.8. Survivorship curves for males of five species of *Desmognathus*. The vertical axis is on a logarithmic scale. (From Organ 1961.)

area had a biennial reproductive cycle, and the difference between the numbers of gravid females and those that were "spent" was used to estimate the difference in survival caused by breeding, on the assumption that nonbreeding females survived as well as males. This interpretation gave the survivorship of females a "stepped" nature.

Clutch size was closely related to body size among the five species, ranging from a mean of 31 eggs in *D. quadramaculatus* to 6.7 in *D. wrighti*. Assuming an equal sex ratio among eggs, Organ was able to construct life tables for the five species. The assumption of stationary populations incorporated net reproductive rates of 1.0, and it is not surprising that the results came close to that value.

Organ's study established much of what is known, but later work by Tilley (1973a,b, 1974, 1977) and myself has revealed unsuspected geographic variations. Tilley showed that in areas farther south than Organ's study site, *D. ochrophaeus* breeds annually. Other variations include the following: Some populations of *D. monticola* apparently remain larvae for only a few weeks (Hairston, personal observation); others overwinter as larvae (Organ 1961). Similar observations have been recorded for *D. ochrophaeus* in different parts of its range. *D. fuscus* larvae appear always to overwinter. This occurs throughout the extensive range of that species. The larvae of *D. quadramaculatus* persist for at least two years, some observations suggesting a three-year larval period (Bruce 1985). At the opposite extreme, *D. wrighti* have no larvae, but hatch as metamorphosed animals.

Synthesis

The different species of salamanders have many different kinds of life histories, from the typically amphibious, with aquatic larvae and terrestrial adults, to permanently aquatic neotenic forms at one extreme and completely terrestrial forms with no larval stage at the other extreme. Most species with amphibious life histories have relatively brief larval periods lasting a few months, but there are exceptions among those living in permanent bodies of water. The greatest variation in the duration of the life history is found among the completely aquatic species. Some of the neotenic ones have long immature periods and long adult lives. The shortest salamander life cycles are found among the species with facultative neoteny. They become mature in less than one year, whether metamorphosing or not, and tend to have lower annual survival rates than species with an obligate terrestrial stage. Among the most successful family, the Plethodontidae, there has been a repeated trend toward a completely terrestrial existence. Regardless of the kind of life history, salamanders have much longer generations than other animals of comparable size. Even those with relatively short life histories live for two or more years, and many species have immature periods of three years or longer.

4

Communities of pond-inhabiting salamanders

Introduction

All salamanders are voracious predators, but the ecological relationships between species of salamanders, and between them and other coexisting species, have been found to be quite different in different habitats. In this chapter, attention is confined to those forms inhabiting still bodies of water. Observations and experiments on terrestrial salamanders and on communities in streams are covered in Chapters 5 and 6, respectively. Ponds are the typical habitat for many species, and it is typically the larval stages that are involved, as the adults of most pond-living salamanders leave the habitat after breeding. There are a few species inhabiting permanent bodies of water that contain predatory fish, but the pond habitat of many aquatic salamanders is temporary, a shallow depression that fills with water from autumn or winter rain and then dries up in early summer or midsummer. There is thus a premium on early breeding, in order to complete development and metamorphosis before the disappearance of the aquatic environment. This has led to a trade-off between the risks of freezing and drought, a trade-off that has been resolved differently by different coexisting species, as described in Chapter 3.

Consideration of ecological interactions among pond-dwelling salamanders begins with competitive relationships. These are discussed from the points of view of habitat partitioning, of density dependence, and of experimental manipulations. As predators, salamanders have been observed to have effects on other members of the community, both on other salamanders and on various other species, including tadpoles and invertebrates. Predation by salamanders will occupy the second part of the chapter. Finally, salamanders have been found to be the prey species in some studies, and discussion of this will close the chapter.

Competitive relationships

Habitat partitioning

Studies reporting different distributions of coexisting salamander species have been made in Poland (Szymura 1974) and the eastern United

States (Anderson & Graham 1967; Worthington 1968, 1969; Heyer 1976). Szymura's study involved four species of newts (*Triturus cristatus*, *T. alpestris*, *T. vulgaris*, and *T. montandoni*). Multispecies populations of larvae were followed in three ponds, each of which was sampled in more than one location several times a year. The relative abundance of adults of all four species was observed in one of the ponds during the spring, when their breeding seasons overlapped. Szymura was of the opinion that the observed differences in spatial distribution, plus the range of size differences among the larvae, operated to reduce competition for food. In a comparison of two ponds it was found that spatial separation of the species was enhanced by increased heterogeneity of the environment within a pond. The findings are consistent with more general statements about the coexistence of newt species in Germany and England (Steward 1969). It should be noted, however, that there is much overlap in the diets of aquatic newts. Avery (1968) examined the stomach contents of 251 larvae and 274 adults of three species (*T. cristatus*, *T. vulgaris*, and *T. helveticus*) from two ponds in the west of England and found close similarity between the types of prey taken by *T. vulgaris* and *T. helveticus*. The larger *T. cristatus* had eaten many of the same items and also had taken large numbers of snails and isopods. Avery recognized that the similarity of food was too great to allow niche partitioning on that axis and hypothesized that the temperature relations defined a better separation. Griffiths (1986) found no partitioning of food by *T. vulgaris* and *T. helveticus* in a pond in Wales and concluded that food is not limiting for those two species.

In the eastern United States, species of *Ambystoma* have attracted attention. Most workers have found ecological differences, but the differences have not been in the same niche dimensions. Anderson and Graham (1967) and Hassinger, Anderson, and Dalrymple (1970) recorded vertical migration at night by the larvae of *A. opacum* and *A. tigrinum*. The larvae remain hidden on the bottom of the pond during the day. They feed in the stratum near the surface for several hours, after which they return to the pond bottom. In the study by Anderson and Graham, another species, *A. jeffersonianum*, also exhibited this behavior, and it was observed that when either it or *A. tigrinum* was in the same pond with *A. opacum*, there was a sequence, with the result that only one species was at the surface at a time. It will be recalled that *A. opacum* lays its eggs in the autumn, before the temporary ponds have filled with water. They hatch as soon as they are inundated. *A. tigrinum* lays its eggs shortly after the ponds fill, and *A. jeffersonianum* follows in late winter. *A. maculatum*, the last to breed, did not show any migratory behavior in the ponds, nor did either the larvae or the aquatic adults of the salamandrid *Notophthalmus viridescens*. Newly hatched larvae do not migrate, but after *A.*

tigrinum became large enough to do so, *A. opacum* larvae shifted their migration to later in the night, when the *A. tigrinum* larvae had begun to settle to the bottom. In ponds where *A. jeffersonianum* coexisted with *A. opacum*, the former did not begin to exhibit vertical migration until the latter were beginning to metamorphose, at which time they stopped feeding and no longer showed the vertical migration. Anderson and Graham (1967, p. 374) concluded that the sequences "may be important in the utilization of the pond ecosystems and in decreasing competition between the species." In contrast to the observations of Anderson and Graham, Branch and Altig (1981) found in Mississippi that *A. maculatum* larvae migrated into the water column and that they and the larvae of *A. opacum* and *A. talpoideum* overlapped widely in regard to the time of migration and in their positions in the water column. Thus, although there were differences, habitat partitioning would be difficult to invoke as the result of interspecific competition in the ponds that they studied. In a similar vein, Brophy (1980) found extensive overlap in diet between *A. tigrinum* and *N. viridescens*, although the former ate more large prey than the latter.

Two studies carried out in Maryland, south of the New Jersey sites of Anderson and colleagues, considered the coexistence of *A. opacum* and *A. maculatum*. Worthington (1968, 1969) collected samples of larvae of these two species, plus those of *N. viridescens*, over the full period of their growth during two seasons. The most complete data are shown in Figure 4.1. The three species hatched sequentially: *A. opacum* hatched at the end of November, *A. maculatum* in mid-April, and *N. viridescens* in mid-May. There was never any overlap in sizes, and Worthington concluded that the sequence, with the accompanying size differences, had evolved to reduce interspecific competition. In the same area, Heyer (1976) made a study of microhabitat partitioning by the two species of *Ambystoma*. He made sweeps with a dip net in three distinct parts of a pond: the surface layer at the grassy edge of the pond, midwater at a somewhat deeper level farther from shore, and at the bottom of the pond. The sweeps were made weekly from early March to late June in two successive years. *A. maculatum* was distributed nonrandomly in both years, with a preponderance of specimens taken in the surface sweeps at the edge of the pond. *A. opacum*, which was much less abundant in the samples, showed no tendency to be concentrated in any of the three microhabitats. Heyer modified Levins's formula (1968) for niche breadth to calculate an index of "microhabitat breadth" and used Horn's formula (1966) to calculate microhabitat overlap. Both species of *Ambystoma* showed wide variations in microhabitat breadth between the two years, but in neither year was there appreciable overlap in their microhabitats. Heyer (1976, p. 11)

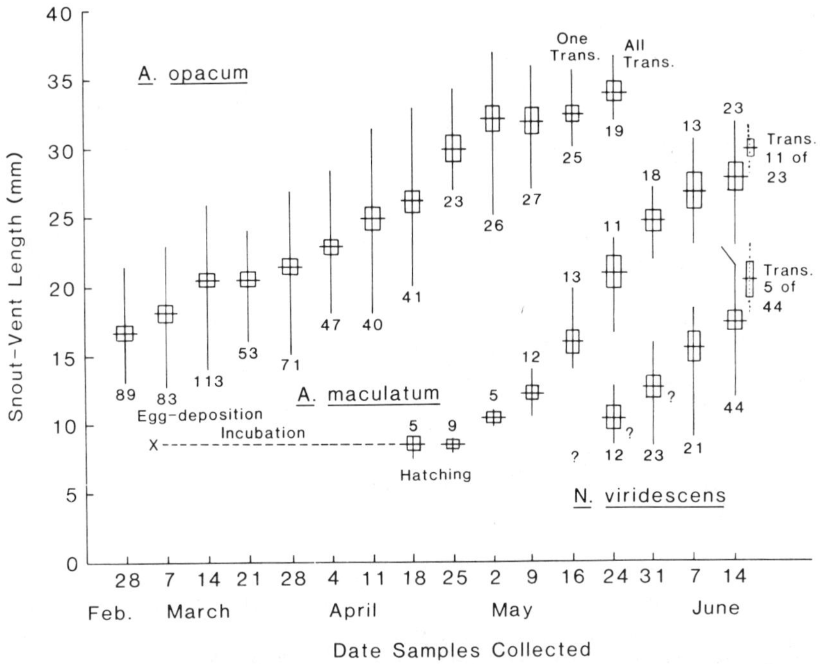

Figure 4.1. Relative sizes of three sympatric species of salamander larvae based on samples collected at weekly intervals from February to June 1965. Numbers indicate number of specimens; vertical lines indicate sample size ranges; horizontal lines indicate sample means; boxes indicate two standard errors of the means; question marks indicate biased samples due to loss of small specimens or a time at which small specimens were predicted. (From Worthington 1968.)

concluded that they "essentially do not co-occur in the same ponds." That is a much stronger statement than any made by other workers, and it is contradicted by the data of Stewart (1956), showing predation by *A. opacum* on *A. maculatum*, and Stenhouse (1984, 1985b) and Stenhouse and associates (1983), showing interactions between the two species in ponds in North Carolina.

Evidence of density dependence among pond salamanders

It is generally agreed that a requisite condition for interspecific competition is a density-dependent effect within the species concerned. That pond salamander populations experience the effect of density has been shown in five studies, two of which were observational, and three experimental. The

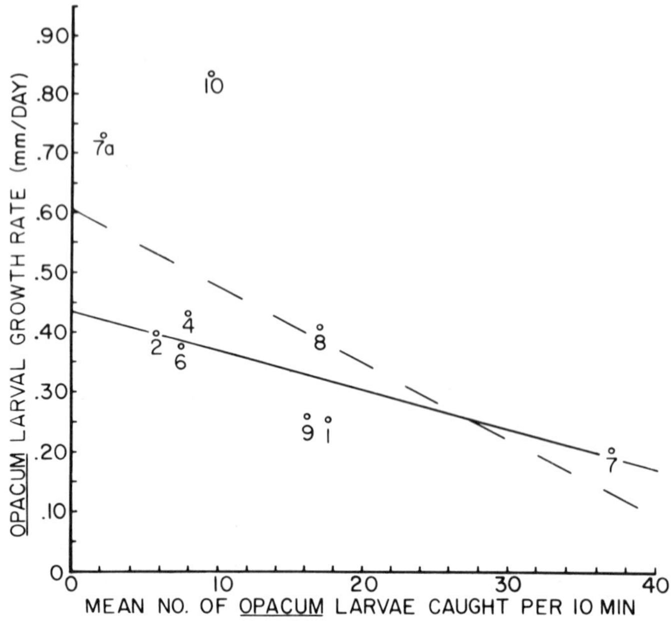

Figure 4.2. Regression of *Ambystoma opacum* growth rate on *A. opacum* density in 1981. Solid line: regression omitting ponds 7a and 10 [growth rate = 0.437 + (−0.00664)density, $r^2 = 0.602$, $p < .025$]. Broken line: regression for all ponds shown [growth rate = 0.606 + (−0.01286)density, $r^2 = 0.382$, $p < .05$]. (From Stenhouse 1985b.)

first report was by Stewart (1956), who observed that for two successive years, growth of *A. opacum* was much slower in a pond containing a high density of larvae than in an apparently richer pond in which the density was low. The second field study was by Stenhouse (1985b), who, like Stewart, worked in piedmont North Carolina. Stenhouse standardized the density of larvae of *A. opacum* in different ponds by averaging the number caught per ten minutes of effort with a dip net. Her study covered nine ponds. There was a statistically significant negative relationship between the growth rate of the larvae and the mean number caught per ten minutes (Fig. 4.2). The reason for calculating the regression in two ways was that ponds 7a and 10 were sampled later in the season than the others and hence were probably affected by higher temperatures. She also found evidence that mortality was higher in ponds where the density was greater.

Three experimental studies in the field, and one in the laboratory, were designed to test for density dependence among pond-living salamanders. Wilbur (1971, 1972) raised *Ambystoma laterale*, *A. tremblayi*, and *A. maculatum* separately in outdoor pens in Michigan. He used two initial densities of each species, 22 and 44/m^2, and recorded survivorship and body weight at metamorphosis, and days from hatching to metamorphosis. The species responded differently to intraspecific crowding. *A. laterale* was appreciably smaller at high density than at low (mean of 0.639 g versus 0.940 g). There were marked differences between replicates that obscured any other effects of density. Neither survival nor length of larval period could be shown to be significantly different between densities, although the responses of larval period to density were shown to be significantly different for the three species. This statistical interaction came about because *A. tremblayi* and *A. laterale* showed no influence of density on survival, but an increase in larval period with density. *A. maculatum*, on the other hand, had few survivors at high density; they performed well and had a shorter larval period than the larger number of survivors at the low initial density. Wilbur conducted simultaneous experiments on competition among the three species, as described later.

The second field experiment testing for the effect of density was that of Gill (1979) on the newt *Notophthalmus viridescens*. In 1974, he selected two ponds in Virginia and in the period March through June transplanted nearly the entire population of one pond into the other. The pond from which the transplanted newts came thus had its population reduced to about 4 percent of the initial density (2,637), and the receiving pond had its population increased to 164 percent of the initial number of 1,713. The results were dramatic. One year after the start of the experiment, both populations had returned to densities close to the original. A considerable proportion of the change was due to newts that homed to the pond from which they had been removed. Most of these were males. Of 536 survivors in 1975, 65 percent (348) were recaptured in the pond from which they had been taken. The overall recapture rate (49.8 percent) was only slightly less than the annual survival of males in control ponds. Recapture of females was only 7.5 percent in both ponds combined; thirty-two had returned to the original pond; twenty remained in the pond to which they had been introduced. All of this movement took place after the autumn exodus of adults from the ponds. Thus, the effects of the manipulation on survival and on body weight could be assessed over the year 1974–5 and compared with the partial-control situation in the same ponds over 1975–6. In July 1974, one month after the manipulations were completed, the mean weights of newts of both sexes in the pond with an increased density were significantly lower than they were at the same time in the succeeding two years. The pond with reduced density gave equivocal results (males heavier than in the two following years, but females lighter). The data for 1975 and

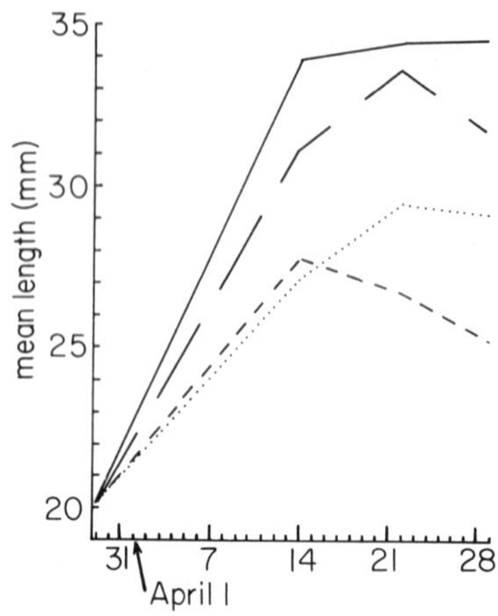

Figure 4.3. Growth of *Ambystoma opacum* in the field competition experiment: Solid line: 17 *A. opacum* alone. Long-dash line: 17 *A. opacum* with 34 *A. maculatum*. Dotted line: 50 *A. opacum* alone. Short-dash line: 50 *A. opacum* with 100 *A. maculatum*. (From Stenhouse et al. 1983.)

1976 were virtually identical for the two ponds. The case for an effect of an increased density on body weight is convincing. Data on survival indicate that for female newts there was an inverse relationship between annual survival and the size of the breeding population in the preceding year. This was also true of the control ponds. The complete experiment shows that the adult newt populations in the ponds are naturally near the carrying capacity and indicates the potential for interspecific competition.

Stenhouse and associates (1983) reported on experiments in small field cages. In one experiment, densities of *Ambystoma opacum* were set so as to be approximately 0.5 and 1.5 times that found naturally. Larvae were photographed with a ruler before being introduced into the cages, and on each of three subsequent times during the next month, as many as could be obtained were dipped from the cages and photographed in the same way. The sizes of the two groups diverged as expected, and at the last measurement, the mean lengths were 34.6 mm and 29.2 mm, respectively – a highly significant difference (Fig. 4.3). Similar experiments, confounded with an attempt to assess the effect of *A. opacum* on *A. maculatum*, and vice versa, gave similar results. In both sets of experiments, the density of *A. opacum* was shown to be most important in reducing growth, but *A. maculatum* also depressed the growth of *A. opacum*, despite the fact that it was preyed on by that species. The interspecific experiments will be described in more detail later. The whole set of experiments shows clearly that there were density effects on growth at densities realistic for the field situation.

The laboratory experiments of Stenhouse and associates (1983) were carried out in round plastic containers 21 cm in diameter. Replicated conditions were of one, two, four, and eight specimens per container. Three sizes of *A. opacum* were used, but only the one size of *A. maculatum* that was available at the start of the experiments. The densities ranged from 0.875 to 7.0 times the natural density for *A. opacum*. Food was provided daily in sufficient quantity to satiate the animals at the two lowest densities and at the sizes prevailing at the start of the experiments. Increased density had a negative effect on the growth of *A. maculatum* larvae and on all three sizes of *A. opacum* larvae. Thus, the results in the better-controlled conditions in the laboratory confirmed findings in the more natural field experiments.

Experiments on interspecific competition among pond salamanders

Wilbur (1971, 1972) carried out detailed experiments on interspecific competition in one of the most complex guilds of *Ambystoma*: a four-species array in southern Michigan consisting of *A. maculatum*, *A. laterale*, *A. tremblayi*, and *A. tigrinum*. The first three species belong to the maculatum group of the genus, and Wilbur's experiments on them progressed from testing the effect of intraspecific density (as described earlier) through the combinations of two species to the whole three-species system. The experiments were carried out in pens of window screening set along the edge of a permanent pond. Each experimental treatment was represented once in each of two blocks. Because of the large variance between replicates, the blocks did not enter into the analysis of the results.

In the experiments in which pairs of the three species were tested for competition, two densities of each (22 and 44/m^2) were used in a completely crossed design. The results were assessed by survival, duration of the larval period, and size at metamorphosis. Except for the experiment assessing competition between *A. laterale* and *A. tremblayi*, differences between replicates were too great to allow the detection of any effects on survival. The higher density of *A. laterale* reduced survival of *A. tremblayi* and also caused an intraspecific effect. The effects of the different treatments on the duration of the larval period and on body weight at metamorphosis were more easily detected, partly because the error mean squares were based on individual survivors to metamorphosis, rather than on the means in replicate pens. Treatment effects were thus confounded with the effect of the location of the pens, but with few exceptions there is little evidence that location was of great importance. This problem is a common one in ecological experiments and is discussed at length by Hurlbert (1984) under the term "pseudoreplication."

The duration of the larval period is important in temporary ponds, because with a long period, development may not be completed before the pond disappears. Important effects of increased density were found in all three experiments involving two species. The effects involved increases in both intraspecific and heterospecific densities. *A. laterale* and *A. tremblayi* each responded to an increase in the other's density, but only when the intraspecific density was high. The same result was found in the response of *A. laterale* to *A. maculatum*, but in the third combination, *A. tremblayi* responded to an increase in the density of *A. maculatum* only at low densities of conspecifics, and *A. maculatum* responded only to an increase in its own density. In the three-way experiment, an increase in the density of *A. maculatum* increased the duration of the larval period for both of the other species, but not the duration of its own larval period. There was an intraspecific effect on *A. laterale*, but not on *A. tremblayi*. The results of the experiments on body weight at metamorphosis were also complex and were not additive as the number of species increased. In the two-way experiments, an increase in the density of each of the three species affected its own final weight. The density of *A. tremblayi* had a significant effect on *A. laterale*, but the reverse was not found. The final weight of *A. tremblayi* was affected by an increase in the initial density of *A. maculatum*. In the three-way experiment, the overall results were affected by the densities of both *A. laterale* and *A. tremblayi*. It was only through its interaction with the other two species that *A. maculatum* was important. The analysis revealed that all of the two-way interactions between the densities of the three species were significant. Wilbur interpreted his results to mean that these significant interactions reflected the fact that the total number of salamanders in the system, rather than the relative abundances, was most important. He considered that the result came from the ecological similarity among the three species.

Despite the similarity, competition in this guild was complicated by the nonadditive effects referred to earlier. This was shown by the fact that doubling the number of competing salamanders did not give consistent results. *A. laterale* showed a sharp decrease in body weight with the addition of either thirty-two *A. maculatum* or thirty-two *A. tremblayi*; further addition of thirty-two more of either had little effect on the weight of *A. laterale*, but the addition of sixty-four of each of the competitors resulted in another sharp drop in the weight of *A. laterale*. There were several other irregularities in the responses of the other two species. These nonadditive effects revealed the existence of "higher-order interactions" among the species. These can be thought of as "coalitions" that either reduced the effect of one of the species on the other two or increased the effect of two species on the third beyond the sum of their separate effects. Wilbur showed that the variance in the final composition of the different combinations of experimental "communities" declined with the increase in the number of species used. He attributed this increased stability to the multiple higher-order interactions among the three species combined.

The fourth local species of *Ambystoma*, *A. tigrinum*, was a predator on the other three, and the experiments using it are described later. As a side result, the addition of tadpoles of the wood frog, *Rana sylvatica*, to the experiment reduced the performance of the maculatum group of salamanders. Because tadpoles are not predators, this was apparently due to their competing with the herbivorous prey of the salamanders. More tadpole–salamander interactions are described later in this chapter.

The existence of a fifth species of *Ambystoma* in ponds not many kilometers from the experimental area prompted Wilbur to carry out one experiment using that species, *A. texanum*. Wilbur added sixteen larvae of that species to replicate pens containing sixteen of each of the cooccurring species. Controls with *A. texanum* alone showed that it was able to grow to metamorphosis in the experimental pond. There were, however, only two survivors in the two replicate pens with the maculatum group, and the other responses (body weight and length of larval life) were also worse than in the controls, although not by a statistically significant amount because of the small number of survivors. By all three responses, the maculatum group performed worse with *A. texanum* than without it, significantly so in the case of body size. Inasmuch as the densities were lower than the main experiment, it appears that competition was intense, and it is doubtful that *A. texanum* could invade a community of the other three species where they were in high densities.

In the piedmont of the southeastern United States, there are only two species of *Ambystoma*: *A. opacum* and *A. maculatum*. As will be shown later, the principal interaction between them is predation by *A. opacum* larvae on those of *A. maculatum*. This comes about because of the difference in times of reproductive activity. By the time that *A. maculatum* has laid its eggs and they begin to hatch, *A. opacum* has hatched and grown to a size at which it can act as a predator on the smaller larvae. Nevertheless, Stenhouse and associates (1983) found that the growth of each species was reduced by the presence of the other. The field experiment described earlier under density dependence also involved pens in which the two species were grown together. As in the "*opacum* only" pens, the initial density of *A. opacum* was set approximately equal to 0.5 and 1.5 times the natural density, as estimated by two different methods. Twice as many larvae of *A. maculatum* as of *A. opacum* were added to the pens. The number of *A. maculatum* added was determined by a crude estimate of their abundance. After one month, the mean lengths of *A. opacum* grown under the different conditions varied by a factor of 1.32 of the mean in the pens with the largest number of larvae (Fig. 4.3). At both densities of *A. opacum*, the mean length of that species was significantly less in the pens with the smaller *A. maculatum* than in the pens with *A. opacum* alone. Even though predation on *A. maculatum* is the predominant interaction in this community, as described later in this chapter, that species survived long enough to exert a significant competitive influence on the larger species. By the end of these field experiments, there were too few larvae of *A. maculatum* left to assess the competitive effect of *A. opacum*. The effect could be shown in the laboratory experiments, however. Those experiments will be described in more detail later, as their primary purpose was to reveal the effect of predation. After three weeks in the same plastic containers, the mean length of *A. maculatum* larvae was 26.8 mm, whereas controls raised at the same intraspecific densities had a mean length of 29.5 mm. The difference is statistically significant.

In the Sand Hills region of the coastal plain of North Carolina, a different combination of salamander species occupies the ponds. The two common species are the newt, *Notophthalmus viridescens*, and the tiger salamander, *Ambystoma tigrinum*. The latter has a typical amphibious life history. The adults come to ponds in early winter and, after mating, the females lay eggs in gelatinous masses of thirteen to fifty-two ova (Morin 1983b) and return to the terrestrial habitat. *Notophthalmus*, on the other hand, remain in the pond as adults, which breed and lay eggs later in the spring or early summer. There is an immature terrestrial stage (the eft) in the life history that begins at metamorphosis and lasts typically for several years. The relationship between the two species involves both competition and predation.

Morin (1983b) manipulated densities of the two species under seminatural conditions in metal tanks 1.52 m in diameter by 0.62 m deep. Food for the salamanders consisted of 1,200 frog tadpoles, plus a generous mixture of aquatic insects and crustaceans. In the present context, the experiment consisted of four replicate tanks with four adult newts each, three tanks with four larvae of *A. tigrinum* each, and three tanks with four newts and four larvae of *A. tigrinum*. The effect of *Notophthalmus* was to reduce the size of *Ambystoma* at metamorphosis to a statistically significant degree and to lengthen the larval period, although the latter effect was not significant. There was no effect on survival. *Ambystoma* had a complex effect on *Notophthalmus*. The mean number of efts emerging from the control tanks without *A. tigrinum* was

101 ± 19; in the presence of *A. tigrinum* the average was 27 ± 16 per tank – a significant difference. Efts compete with each other and with adult newts. The newt larvae at reduced densities in the presence of *A. tigrinum* transformed into efts that were twice as large as those in the control tanks (a significant difference), and the adult newts in the control tanks were significantly smaller than in the *Ambystoma* tanks, where the efts were fewer in number. Morin advanced the explanation that predation by larval *Ambystoma* reduced the number of efts in their tank, thereby reducing intraspecific competition among the adult newts and the efts.

In combination with the observations and experiments showing density dependence among the salamanders in the pond community, these experiments demonstrate the extent of interspecific competition in these salamander guilds. The competitive relationship is complicated by asymmetrical influences among the competitors, and the situation in many pond communities is further complicated by predation by some members of the guild on others. A consideration of predation follows.

Predation by salamanders in the pond community

Field observations

Salamanders swallow their prey whole, a habit that encourages examination of their stomach contents. Provided that a minimum of time has elapsed since feeding, most of the food can be identified. A short interval is important, because prey with chitinous parts remain intact much longer than prey that are soft-bodied, which are digested rapidly to an unrecognizable state. As far as pond-dwelling salamanders are concerned, there has been a limited number of analyses of stomach contents. The studies of Avery (1968) and Griffiths (1986) were described earlier under habitat partitioning. In the present context, predation by salamanders will be described for field situations where they have been found preying on other salamanders or have been shown to have had a sufficient predatory effect to alter the composition of the prey community.

Predation among pond-living salamanders is common when there is a sufficient size difference. Such a situation occurs among early-breeding and late-breeding species. As the species coming last to breeding ponds, *Ambystoma maculatum* is most frequently the prey. The eggs hatch after the larvae of the coexisting congeners have grown to a size large enough to prey on those of *A. maculatum*. The species most often observed to be the predator is *A. tigrinum*, which breeds early and grows rapidly. Predation has been observed among

Table 4.1. *Predation between larvae of seven species of* Ambystoma *in the eastern United States; the combinations in which an effect was observed are documented by references*

	Predator species			
Prey species	*laterale*	*tigrinum*	*texanum*	*opacum*
laterale	Bishop (1941)	Wilbur (1972)		
tremblayi		Wilbur (1972)		
jeffersonianum				Walters (1975)
texanum		Hay (1889)	Hay (1889)	Walters (1975)
maculatum	Bishop (1941)	Bishop (1941)		Stewart (1956)
		Wilbur (1972)		Stenhouse et al. (1983)

seven species occurring in the eastern United States (Table 4.1). Table 4.1 includes results of experiments as well as results from examinations of stomach contents of specimens caught in the field. In addition to the observations cited in Table 4.1, Morin (1983b) found presumptive evidence that *A. tigrinum* larvae prey on the larvae of *N. viridescens*.

Predation on other species is not confined to their larvae. Three observers have demonstrated predation by salamander larvae or adults on the eggs of other salamanders, and in two cases they thought that the effect on the prey population was powerful enough to eliminate it in some years. Morin (1983b) observed adults of *N. viridescens* attacking the egg masses of *A. tigrinum* in a pond in North Carolina and was able to recover no metamorphosed individuals of the latter at the end of the season. Walters (1975) reported similar observations on the effect of predation by *A. opacum* on the eggs and larvae of *A. texanum* and *A. jeffersonianum*. Her observations included years in which *A. opacum* did and did not breed successfully in the same pond, and her conclusion was thus convincing, as the difference was decisive for the success of *A. texanum*. Although he did not observe the effect on the population of *A. maculatum*, Hamilton (1932, p. 84) found that its eggs were "the important food" of *N. viridescens* during late March.

Salamander predation may have important effects on the species composition of the invertebrate part of pond communities. Dodson (1970) and Dodson and Dodson (1971) investigated zooplankton communities in twenty-four similar ponds in the mountains of Colorado and found a striking dichotomy in taxonomic composition that was paralleled by a marked size difference. Ponds that were inhabited by the larvae of *A. tigrinum* contained only small

Table 4.2. *Relative frequency of* Daphnia pulex *as percentage of total* Daphnia, *Leechmere, 1968*

Date	In plankton (%)	In salamander diet (%)
9 July [a]	38	35
19 July	11	38
29 July	11	19
8 Aug.	0.5	5

[a]The salamander eggs hatched on about this date.
Source: Dodson (1970).

microcrustaceans, the copepod *Diaptomus coloradensis* (0.1–1.2 mm long) and the cladoceran *Daphnia rosea* (0.4–1.5 mm). Ponds without the salamander larvae supported crustacean communities consisting of the fairy shrimp, *Branchinecta coloradensis* (0.2–12 mm), *Daphnia pulex* (0.7–3.0 mm), and the copepods *Diaptomus shoshone* (0.2–3.0 mm) and *D. coloradensis* (the only ubiquitous species). In some ponds the salamanders metamorphosed in late summer and left. In those ponds, the zooplankton community, early in the season, before the new crop of salamanders hatched, consisted of the large species. As soon as the salamander larvae appeared (around July 9), there was a rapid change to the small species of zooplankters. The conclusion that the cause was predation on the large forms is supported by Table 4.2, showing the relationship between the representation of *D. pulex* in the plankton and in the stomach contents of *A. tigrinum* larvae. In other ponds the salamanders overwintered as larvae, and there was a period of about a month when both generations were present. The stomach contents of the larvae showed a distinct preference for the larger size classes of prey. This size-selective predation clearly determined which zooplankton community would be present.

Experiments on predation by salamanders

Several of the field observations described earlier were sufficiently detailed to give convincing evidence that predation by salamanders had an important influence on the prey communities. Many of them were anecdotal, however, and under such circumstances it is important to have experimental

Table 4.3. *Effects of* Ambystoma tigrinum (A.t.) *and* Rana sylvatica *on the maculatum group: data summary*

Species	Initial density per pen	Survival (%)		Mean larval period (days)[a]		Mean body weight (g)	
		No. of *A.t.*		No. of *A.t.*		No. of *A.t.*	
	R. sylvatica	0	5	0	5	0	5
A. laterale	0	69	59	94(22)	76(19)	0.814	0.661
	300	59	3	98(19)	126(1)	0.746	0.793
A. tremblayi	0	56	38	74(18)	74(12)	0.826	0.771
	300	38	3	93(12)	92(1)	0.773	1.403
A. maculatum	0	50	28	77(16)	86(9)	1.086	1.061
	300	72	22	110(23)	107(7)	0.940	1.328

[a]Numbers of survivors from both blocks are in parentheses; all species of the maculatum group had an initial density of sixteen hatchlings per pen.
Source: Wilbur (1972).

demonstrations that a significant impact was involved. After descriptions of experiments on the effects of salamander predation on other salamanders, there are descriptions of experiments demonstrating the effects of salamander predation on other members of the pond community.

As part of his experimental analysis of the *Ambystoma* community in southern Michigan, Wilbur (1972) assessed the effect of *A. tigrinum* on the members of the maculatum group (*A. laterale*, *A. tremblayi*, and *A. maculatum*). The experiment was combined with a test of the effect of the tadpoles of *Rana sylvatica* (Table 4.3). Whether the tadpoles were present or not, the larvae of *A. tigrinum* reduced the survival of all three species of its congeners. A study of the table reveals the complexity of the combined effects of *A. tigrinum* and *R. sylvatica* on the maculatum group. As mentioned earlier in regard to competition, the presence of the tadpoles caused a prolongation of larval life; the predator did not. The presence of *A. tigrinum* larvae resulted in an increase in body size for each member of the maculatum group, but only when the *R. sylvatica* tadpoles were present. That result suggests that the predation on both the maculatum group and the tadpoles resulted in less competition for the surviving salamander larvae, allowing them to reach a larger size at metamorphosis. Thus, in addition to the complex competitive interactions, the presence of a predator on the competing species adds further to the difficulty of constructing a mathematical community theory of a reasonable degree of generality.

A different *Ambystoma* community was investigated by Stenhouse (1984, 1985b) and Stenhouse and associates (1983). *A. opacum* lays eggs in the autumn, where pools will appear later, and they hatch when covered with water by the winter rains. During the winter, the larvae grow slowly, but still reach a total length of at least 29 mm (mean = 33 mm) by late March – the normal time of hatching for *A. maculatum* at a length of 13 mm. The size difference is ample to allow predation by *A. opacum* on the smaller larvae. In the piedmont of North Carolina, there are few ponds with *A. maculatum* that do not also contain *A. opacum*. Stenhouse and associates (1983) showed in laboratory experiments that even small larvae (initial mean length = 23.8 mm) of *A. opacum* preyed extensively on their smaller congeners. In ponds where both species are present, the effect of this predation on the population of *A. maculatum* is worthy of investigation, because in many years, no metamorphosed individuals leave the pond. Stenhouse (1984), for example, monitored a pair of ponds together for two years. Careful counts of the number of egg masses of *A. maculatum*, and accurate calculation of the mean number of eggs per mass, allowed her to estimate the proportion that were trapped at the drift fence when they left the ponds after metamorphosis. In 1980, she estimated that 3,672 eggs were laid; four juveniles were intercepted leaving the ponds – a survival rate of 0.0011. In the next year, there were none. *A. opacum* had much better success, 230 juveniles having left the pond from 2,607 eggs – a survival of 0.0882, eighty times as great as for *A. maculatum*. In 1980, the eggs of *A. maculatum* were produced by a total of seventy-one adults. Stenhouse calculated that if all juveniles survived until recruitment balanced losses in the pond, they would have to live 17.75 years – an unrealistic figure.

The study was undertaken to assess the effect of predation by *A. opacum*. It is important to know the amount necessary to prevent successful recruitment by *A. maculatum*, because it must be a temporary condition, individuals of both species being faithful to a pond. Alternatively, migration by juveniles from ponds that are successful might maintain a population in unsuccessful ponds, as the adults are long-lived (see Chapter 2). Knowledge of movement among ponds is confined to anecdotal information. Stenhouse and I have observed eggs of *A. maculatum* in newly created pools, an observation showing that some movement among ponds must take place, but its magnitude remains unknown. Differences among ponds in the impact of predation could be caused by differences in numbers of *A. opacum*, or there could be better refuges or alternative prey in some ponds. Stenhouse investigated these possibilities. The density of *A. opacum* larvae early in the spring varied among ponds by a factor of nearly thirteen (Fig. 4.2). That finding suggests that

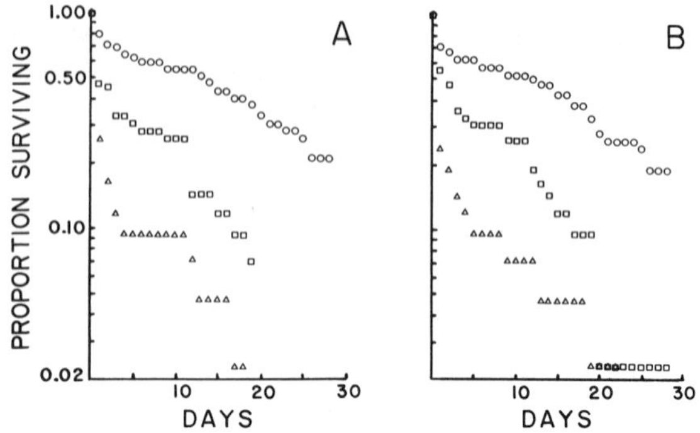

Figure 4.4. (A) Survival of *Ambystoma maculatum* larvae in the laboratory with "small" *A. opacum* (circles), "medium" *A. opacum* (squares), and "large" *A. opacum* (triangles). (B) Survival of *A. maculatum* larvae in the laboratory with one *A. opacum* (circles), two *A. opacum* (squares), and four *A. opacum* (triangles). (From Stenhouse et al. 1983.)

predation on *A. maculatum* may vary comparably, providing some ponds with successful recruitment for that species.

Laboratory experiments showed that this is a realistic possibility. All combinations of one, two, and four specimens of both *A. maculatum* and *A. opacum* were kept together in plastic dishes 21 cm in diameter, with enough food (tubificid worms) to satiate two *A. opacum* larvae. The experiment was performed with three sizes of *A. opacum*. The death rate of *A. maculatum*, all from predation, was significantly greater with increasing densities of *A. opacum* (Fig. 4.4B). The daily mortality was 0.059 with one *A. opacum*, 0.134 with two, and 0.170 with four. These calculations include three initial sizes of *A. opacum*. In a single pond, two different methods agreed in an estimated density of thirty-three larvae of *A. opacum* per square meter at the time when *A. maculatum* was hatching. That is slightly higher than the lowest density (28.9/m^2) in the plastic dishes. If there were no other cause of mortality, the expected proportion of eggs becoming metamorphosed juveniles would be 0.029 for a sixty-day period when both species were in the pond. At the intermediate experimental density, the expected survival over sixty days, and with no other source of mortality, would be 0.0003. Unfortunately, we do not know whether the observed density was typical or extreme. The fact that the success rate of *A. maculatum* was only 0.0011 in that season suggests that it might have been fairly high – a suggestion that agrees with a subjective assessment (Stenhouse 1984, p. 60). Stenhouse and associates (1983) also carried out field experiments on the effect of *A. opacum* on the survival of *A.*

maculatum. The densities in circular cages 0.5 m^2 in area were ten and twenty *A. opacum* and twenty and forty *A. maculatum*. The predator was at 0.61 and 1.22 times the observed natural density. The cages had natural clay bottoms, covered with leaves, as was the pond bottom, thus providing some cover. The experiment was terminated after twenty-three days. The daily mortality rate for *A. maculatum* in control pens (forty per pen) was 0.074; with twenty *A. opacum* it was 0.195; and with ten *A. opacum* it was 0.162. Comparable mortality in the laboratory experiments was achieved only at nearly six times the density of *A. opacum*. Stenhouse and associates interpreted that to mean that food in the field pens was much less available than in the laboratory, causing the *A. opacum* larvae to prey more specifically on *A. maculatum*. A second experiment with larger pens (1.47 m^2) appeared to confirm the interpretation. The density of *A. opacum* larvae was set at 3.4/m^2 and 27.2/m^2 in two sets of pens that were replicated three times. The density of *A. maculatum* larvae was 34/m^2. There were controls with no predators. Survival was followed until metamorphosis, the period including sixty-six days when both species were present. The daily mortality for *A. maculatum* was only 0.036 in the control pens, 0.031 in the pens with the low density of *A. opacum*, and 0.052 in the high-density pens. The low density of the predator had no demonstrable effect on the mortality for the prey.

In a series of laboratory experiments, Stenhouse (1984) showed that survival of *A. maculatum* was significantly improved when there was a refuge available in the form of aquatic vegetation (*Potamogeton* species) or dead leaves. The presence of alternative prey (tadpoles of *Bufo americanus* or *Pseudacris triseriata*) also increased the survival of *A. maculatum* larvae. There are therefore several ways by which *A. maculatum* achieves reproductive success, even though it appears to be exposed to possibly severe predation by *A. opacum* in all of the ponds in which it breeds.

The relationship between *Notophthalmus viridescens* and *Ambystoma tigrinum* in the Sand Hills region of the North Carolina coastal plain was described earlier in regard to competition. Morin (1983b) observed that adult newts were able to work their way into the egg masses of *Ambystoma* and prey extensively on the eggs. By protecting some egg masses from newts in a natural pond, he showed that without such protection no *Ambystoma* larvae were produced at all in some years. In other ponds with fewer newts, they coexisted with larval tiger salamanders, which grow rapidly to a size large enough to escape predation. Under these conditions, the relationship changes to one of competition, but not completely so, as the larvae of *A. tigrinum* rapidly grow to a size at which they prey on the eggs and larvae of *N. viridescens*. Adult newts are distasteful and apparently immune to the predation, despite being in the size range that would otherwise be vulnerable. In the experimental tanks containing realistic numbers of both species, there were produced only about one-fourth as many metamorphosed newts as in tanks without *Ambystoma*. Mutual predation between the two species adds a special twist to their community.

Experimental studies on the effect of predation on frog tadpoles

Salamander larvae have been shown to exert a powerful effect on the guild of frog tadpoles on which they prey. A series of elegant experiments using adult newts and larval tiger salamanders was performed by Wilbur and his students (Morin 1983a; Morin, Wilbur, & Harris 1983; Wilbur, Morin, & Harris 1983). The experiments were carried out in the metal tanks that have already been described. Preliminary surveys established the density and time of breeding of each species. Before the experiments began, the tanks were filled with approximately 1 m^3 of water and supplied with trout chow, dry litter, and stems of the macrophyte *Myriophyllum pinnatum*. During the next three weeks, aquatic organisms known to be natural prey of salamanders were added: plankton, plus amphipods and aquatic insects (Corixidae). Thus, the experiment can best be described as quasi-natural. At the end of the three weeks, predatory salamanders were distributed among twenty-two tanks as follows: four tanks with no adult newts, four with two newts each, four with four, and four with eight; three tanks with four larvae of *A. tigrinum*, and three with four adult newts plus four *Ambystoma* larvae. A total of 1,200 tadpoles in each tank completed the experimental design. They were added as follows: 100 *Rana sphenocephala*, 200 *Scaphiopus holbrooki*, and 300 *Hyla crucifer* at the time of introduction of the predators (27 March), 300 *Bufo terrestris* on 2 May, and 150 *Hyla chrysocelis* plus 150 *Hyla gratiosa* on 24 May. The addition of tadpoles corresponded to the times of their appearance in nature.

The effect of *A. tigrinum* larvae was almost absolute exclusion of tadpoles: only thirteen succeeded in metamorphosing in the six tanks, and the four newts had no detectable additional effect. The experiment with differing densities of newts gave extremely interesting results. There was a nearly linear decline in numbers of surviving anurans with the increase in newt density, but much more interesting was the effect on competition among the species of tadpoles. Without predation, the emerging population was dominated by *Scaphiopus* and *Bufo*. They declined in relative importance as the number of newts increased, and *Hyla crucifer* showed a regular increase in importance, coming to dominate the community when eight newts were pres-

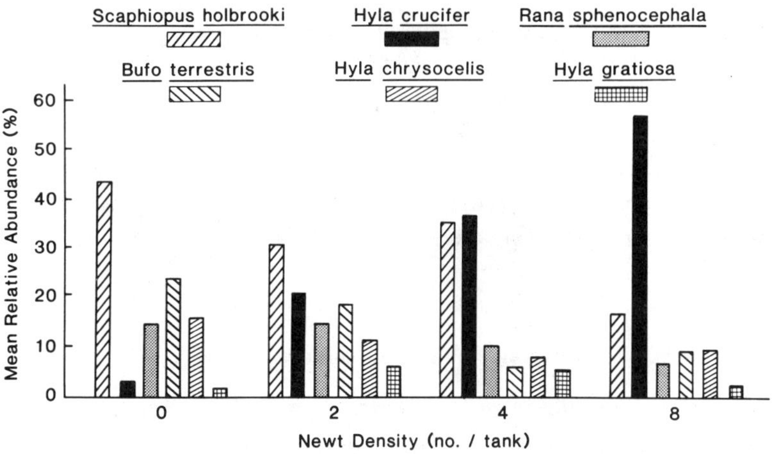

Figure 4.5. Mean relative abundances of surviving censused anurans at each level of *Notophthalmus* density. Relative abundances of *Rana* include combined abundances of metamorphs and overwintering tadpoles. Histograms are identified by the key at the top. All means are based on four replicate communities at each level of *Notophthalmus* density. (From Morin 1983a.)

ent. *Hyla gratiosa* had a negligible representation at the extremes of predation, but performed somewhat better at intermediate densities of newts. The remaining two species declined in importance as the density of newts increased (Fig. 4.5). Realistic levels of predation clearly had important effects on thc composition of the competitive community. Survival to metamorphosis was not the only effect of predation. As increased predation reduced the impact of competition, the size of the tadpoles at metamorphosis increased. The final effect was that the highest level of frog biomass was obtained in the tanks with two newts, and the next highest was in the tanks with four newts. It is widely agreed that larger size increases survival, and there are many studies showing a positive correlation between fecundity and size. The effect of predation was not only to alter the relative abundance of the species of tadpoles at metamorphosis but also to affect the relative postmetamorphosis performance of the species.

In subsequent experiments in the same type of containers on the combined effects of tadpole density and newt density, Wilbur and associates (1983) and Morin and associates (1983) found even more drastic effects on the total abundance of tadpoles than Morin (1983a) had observed. As in his experiment, the surviving metamorphs were much larger than those emerging from the tanks with no predators. These authors also monitored the species of zooplankton during the course of the experiment. There were complex interactions among the species. Tadpole abundance had no effect on the abundance of zooplankters, but newts had interesting and some unexpected effects. The first effect was an increase in the total numbers because of a great increase in very small species. Statistical analysis of the relative abundance of the different species of zooplankton showed significant effects of newt density during the last half of the three-month experiment. The results are much more difficult to interpret than are the effects of predation on the tadpole community.

Predation on pond-inhabiting salamanders

It is axiomatic in eastern North America that the only salamander species in ponds with fish is *Notophthalmus viridescens*. The adults have poison glands in the skin, and in many experimental tests they have been rejected by a variety of potential predators (Brodie 1968; Hurlbert 1970; Formanowicz & Brodie 1982). *Ambystoma* larvae are relatively palatable, and in addition to predation by other salamanders, as has already been described, they have been eaten in experiments by dytiscid beetle larvae, giant water bugs, crayfish (Formanowicz & Brodie 1982), and turtles (Walters 1975).

Two field studies have assessed the effect of fish predation on *Ambystoma* populations. Efford and Mathias (1969) studied two species of salamanders in a small lake in British Columbia. *Taricha granulosa*, like its eastern relative *Notophthalmus*, was immune to predation by the rainbow trout, *Salmo gairdneri*, and the other salmonid *Oncorhynchus nerka*. *Ambystoma gracile* was almost completely neotenic. It was found frequently in the stomachs of the rainbow trout and, unlike eastern congeners, was able to maintain a density of 175 per hectare throughout the lake, and 1,678 per hectare in some areas near the shore. Unlike *Taricha*, the *Ambystoma* were difficult to catch and hid quickly behind a cloud of mud when approached. The dynamics of the salamander populations were quite different. *A. gracile* grows to adult size in one season and breeds every year. Efford and Mathias estimated a life of about five years. Its growth responds to both temperature and food supply. The picture is of a species cropped regularly by trout predation, and Efford and Mathias give the impression that predation is the limiting factor. In contrast, *Taricha granulosa* living in the same lake, and using the same food, grows slowly, probably takes four or five years to reach maturity, and lives for an estimated average of twelve years. Females appear to breed in alternate years. These are features of a stationary population that does not receive perturbations.

The other study in which fish predation was found to be important to a salamander population is that of Petranka (1983). In streams in central Kentucky, he found that there was a strong negative relationship between the distribution of fish and the larvae of *Ambystoma texanum*. Stream breeding is unusual in this genus and, in the area of interest, *A. texanum* has colonized the headwaters of streams that become series of intermittent pools by late May or early June, the pools normally drying completely by late June. The salamanders lay their eggs in winter, and the larvae have time to complete development through metamorphosis before the pools dry completely. The drying precludes permanent occupancy of these headwaters by fish, which may colonize some of them from farm ponds during winter rains. Petranka searched ninety streams for the eggs and larvae of *A. texanum* and for fish. All were easily observed in the clear water. He found eggs or larvae in fifty-one (70 percent) of the first-order streams and nine (75 percent) of the second-order streams that lacked fish. The only overlap in distribution was in one of fourteen second-order streams that contained fish. Laboratory observations showed that six of the local species of fish ate the salamander larvae readily. Over a three-year period, he was able to locate three salamander breeding pools that contained fish in at least one year. By late April in years when fish were present, the density of larvae was only 6 percent of that when fish were absent; none survived to metamorphosis, in contrast to fishless pools, where 3.8 percent, 8.2 percent, and 7.2 percent in three pools completed development.

Synthesis

It is an unspoken assumption among many workers that the regulation of populations of pond-inhabiting salamanders occurs in the larval stage. That may well be because it has not been feasible to study the terrestrial adults, which remain underground except during the breeding season. Work has been exclusively on the larvae. There is much evidence for density-dependent phenomena in ponds. Correlative evidence of habitat partitioning has been claimed in a number of instances, but there is little agreement about what aspect of the habitat is being partitioned: stratum in the water column, time of feeding, nonoverlap of sizes due to times of breeding, and location within the pond have all been claimed, sometimes for the same combination of species. It seems likely that the interpretation of such observations was strongly influenced by the prevailing theory of community organization, which required some form of resource partitioning. Better evidence for density dependence came from experiments in which the numbers of individuals were manipulated, and anticipated responses were monitored.

Density dependence is a prerequisite for interspecific competition, and experiments testing for competition have revealed that it is present in the three communities where the experiments were performed. The nature of the interactions in these communities, however, is such that conventional community theory could not have predicted the structure and dynamics of the communities. The reasons for the difficulty are, first, that the competition in one of them was shown to involve higher-order interactions not envisioned in the theory, and in all three communities, predation by one species on one or more of the coexisting species made the situations unusually complex.

The importance of pond-inhabiting salamander larvae on community interactions is as great between them and the species at other trophic levels as it is among the salamander species themselves. There is strong evidence that in some areas, predation by salamander larvae determines the composition of the zooplankton community by consuming the larger and competitively superior species. Equally interesting is the differential predation by salamanders on the coexisting community of frog tadpoles. In the absence of salamanders, there is intense competition between the species of tadpoles. Predation alters the success by selective predation on the most successful species. The reduction of competition increased the growth of surviving tadpoles to the extent that the total biomass of metamorphs was greater with predation than without it. Finally, there are examples of the influence of fish predation on the population dynamics and distribution of some pond-dwelling salamanders.

At the level of the question of the importance of density dependence, pond-inhabiting salamanders provide strong support for the need for community theory, because interspecific interactions have been found in all experiments. At the same time, the complexities of these interactions have raised a strong challenge to community theory, which has thus far not been able to produce convincing models predicting the kinds of community structure required by the experimental analyses that have been carried out.

5

Communities of terrestrial salamanders

Introduction

More than half of all species of salamanders are purely terrestrial, an anomaly that has been described in Chapter 3. It is even more anomalous that nearly all of these terrestrial salamanders lack lungs – a heritage of the probable origin of the Family Plethodontidae in mountain streams. With only cutaneous and buccopharyngeal respiration available, and the necessity therefore to keep the skin moist, terrestrial salamanders are limited in distribution to environments that provide microhabitats with saturated atmospheres.

Interspecific ecological relationships among these species have been studied most completely for members of the genus *Plethodon*, and in all cases the relationship has been found to be competition, or it has been assumed that competition was the expected relationship. There have been isolated observations of predation on *Plethodon*, but no author has treated seriously the possibility that populations of these salamanders are affected by predation. Dunn (1926) implied that competition was important to members of the genus when he pointed out the size relationships among coexisting species, but it was not until my study of the ecological distribution of salamanders in the Black Mountains of North Carolina that competition between two species was proposed as the most important factor in determining their altitudinal distributions (Hairston 1949).

Plethodon consists of four distinct groups of species, three of which have received attention as having potentially competing pairs of species. The "large plethodons" are confined to the eastern United States, where there are also two groups of "small plethodons." The "western plethodons" are quite distinct from the eastern forms and are biochemically more like another genus, *Aneides*. I first describe my studies on two species of large plethodons in the southern Appalachians. Then I describe Jaeger's experiments on small species. They are the two most complete studies on competition among terrestrial salamanders. The chapter ends with descriptions of work on other communities. In some cases, no competition was found; in others, it was concluded that the observations indicated niche partitioning.

Competition and altitudinal distribution

Observational evidence

When I began my study of ecological distributions in 1940, the prevailing view was that the Appalachians were not high enough for the salamanders to show the kind of altitudinal separation that Schmidt (1936) had found on a volcano in Guatemala. The opinion was based in part on the assembled records of collectors, especially those of Pope (1928) and Bailey (1937). They had reported that their records of the closely related species *Plethodon glutinosus* and the forms later included in *P. jordani* overlapped altitudinally by 793 m. The records, however, were assembled from different locations around the southern Appalachians.

My method was to carry out vertical transects, collecting at each 30.5 m (100 ft) of elevation. The first such transects were made in the Black Mountains of North Carolina, on slopes facing north, south, east, and northwest. Although it was not possible to find *Plethodon* at every elevation, the overall pattern was clear (Fig. 5.1). The maximum altitudinal overlap on any transect was 61 m, although for the Black Mountains as a whole, the overlap was more than 518 m. The two species were found together much less than the total overlap suggested. The *Plethodon metcalfi* of the figure has since been shown to belong within *P. jordani*, because intergradation is found wherever the local high-altitude forms come together. The altitudinal distribution of the third species, *P. yonahlossee*, was not discussed. It is considered in Chapter 8. I proposed that strong interspecific competition would best account for the altitudinal separation of *P. glutinosus* and *P. jordani*. Their overlap did not coincide with any vegetational or other readily identifiable environmental boundary. The discovery that the two species were separated altitudinally in the Black Mountains prompted a further exploration of other mountain ranges in the southern Appalachians, because they had been found to support other morphologically distinct forms of the high-altitude *P. jordani*, and it was of interest to discover if the same ecological situation would be found in the other mountains. I found that the pattern was not consistent (Hairston 1951). In the Great Smoky Mountains, the situation was almost exactly like that in the Black Mountains, but in the Balsam Mountains, which are geographically between the other two ranges (Fig. 5.2), *P. glutinosus* and *P. jordani* overlapped on two transects by 457 m and 595 m. A third transect, near the Balsams on the Blue Ridge of South Carolina, showed complete overlap, with *P. glutinosus* actually found higher than *P. jordani*. If the original interpretation was correct, that competition was the reason for the altitudinal separation

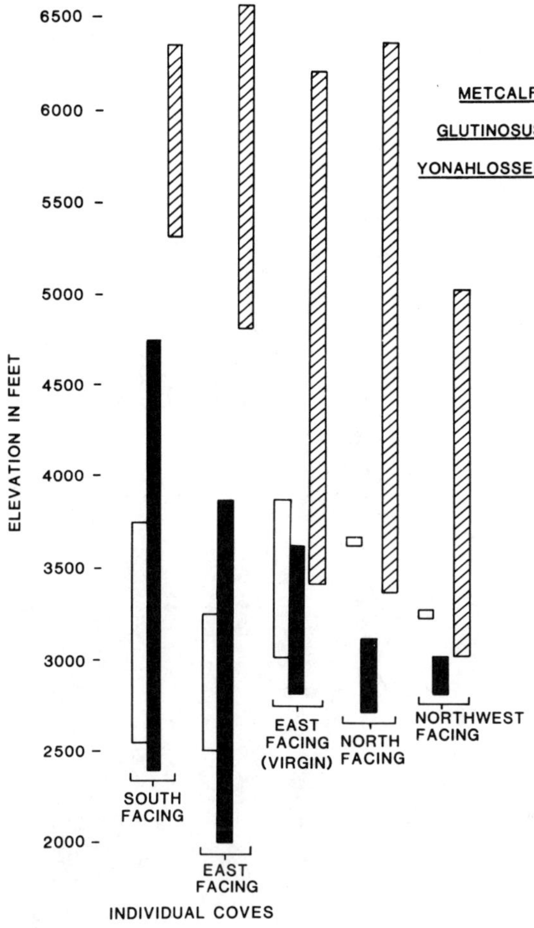

Figure 5.1. Vertical distribution of three species of *Plethodon* in the Black Mountains area, based on observations from transects in individual coves. (From Hairston 1949.)

in the Black Mountains, it must be true that competition was much less intense in the Balsam Mountains. At the time, I thought that the difference was a reflection of geographic variation in the morphologically polytypic *P. jordani*. Subsequent to my studies, a much more extensive investigation was made by Highton (1962a, 1972) and Highton and Henry (1970). My original description of the altitudinal distributions was confirmed and much detail added, including data on electrophoretically distinguishable plasma proteins.

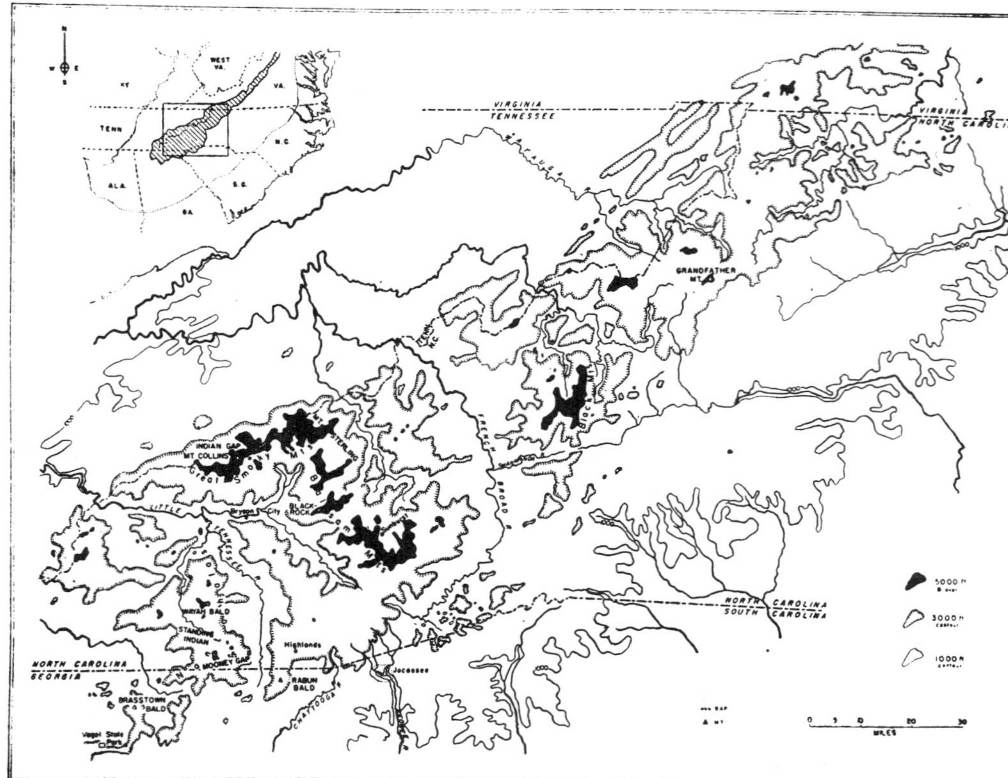

Figure 5.2. Contour map of the southern Appalachian Highlands. Corresponding area is outlined on inset map indicating Blue Ridge Physiographic Province (hatched) and limits of Appalachian Highlands (dotted line). (From Hairston & Pope 1948.)

Experimental tests for competition

In the course of night surveys while accompanying classes to the region, I became convinced that the salamanders were abundant enough to permit field experiments, and I proposed a set of such experiments to test the hypothesis that interspecific competition is much more intense in the Great Smoky Mountains than in the Balsam Mountains (Hairston 1973). The detailed experimental design that was presented made the test of the hypothesis truly a priori, and avoided the circularity that had been rife in community ecology prior to 1973, with the exception of work in the intertidal community. The proposed experiments included reciprocal transplants of *P. jordani*, designed to test hypotheses about evolution under interspecific competition. Those experiments are described in Chapter 7.

The tests of competition involved establishing field plots in the two mountain ranges under conditions as nearly the same as possible: same elevation, same compass direction of slope, and same forest type. The plots, exactly like the one described in Chapter 2, were large enough to contain about forty home ranges of male *P. jordani* or thirty-two home ranges of male *P. glutinosus*, or more home ranges of females or juveniles of either species, based on the then-current estimates of Merchant (1972). On each mountain range, two plots were chosen at random for persistent removal of *P. jordani*, two for removal of *P. glutinosus*, and three to remain as unmanipulated controls. Persistent removals were necessary because only about 25 percent of the specimens are above ground during one search, and because of immigration onto the unfenced plots. The hypothesis of strong competition in the Great Smoky Mountains and weaker competition in the Balsam Mountains predicted that the response of each species to the removal of the other would be more marked in the Smokies than in the Balsams. The experiment was continued for five years, the minimum period of a generation of *P. glutinosus* (Highton 1962b).

Despite an anxious first two seasons (Fig. 5.3), the experiment was a success (Hairston 1980a). In the third, fourth, and fifth years, *P. glutinosus* in the Great Smoky Mountains increased significantly above controls on the plots from which *P. jordani* was removed, demonstrating a negative effect of *P. jordani* on its larger and less abundant congener. The removal of *P. glutinosus* did not result in a significant increase in the abundance of *P. jordani*, but it did result in a highly significant increase in the proportion of the two youngest age classes, indicating that reproduction by *P. jordani* was more successful where *P. glutinosus* was removed. In the Balsam Mountains, there was a positive response by *P. glutinosus* to the removal of *P. jordani*,

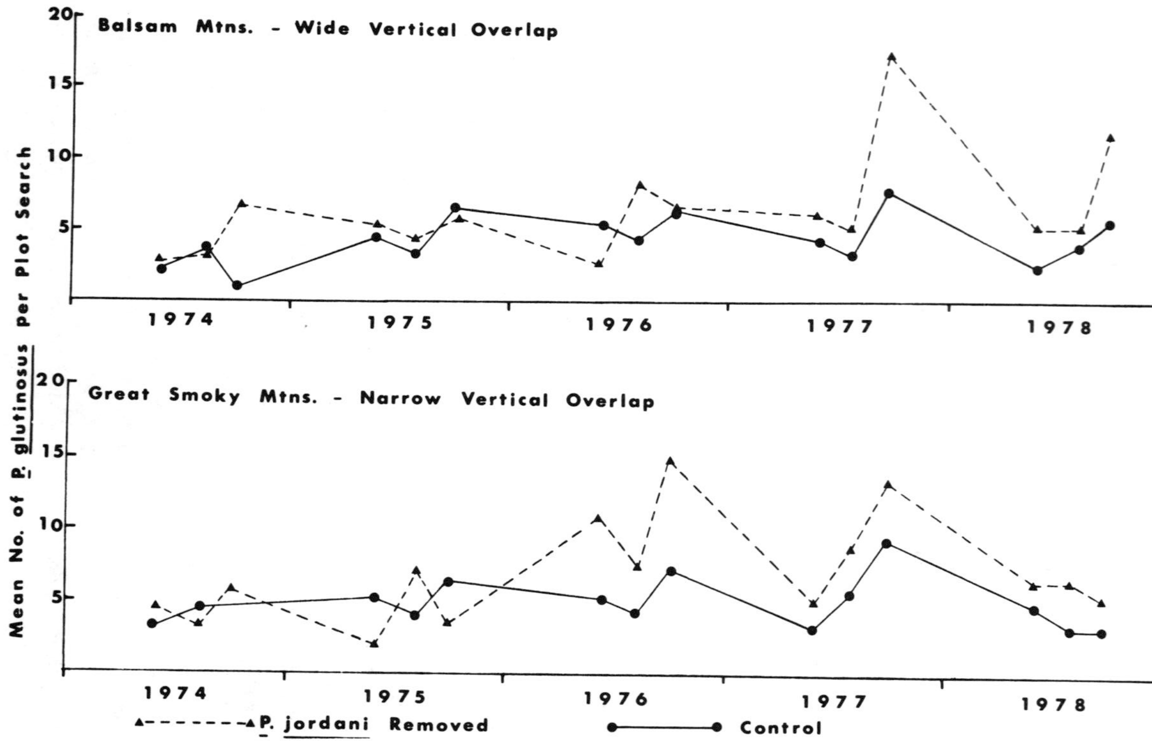

Figure 5.3. Average number of *P. glutinosus* per plot search on plots from which *P. jordani* was removed, compared with the density on control plots. Experimental results from both Great Smoky Mountains and Balsam Mountains are shown. (From Hairston 1980a.)

Table 5.1. *Coefficients of competition between* Plethodon jordani *and* P. glutinosus, *and the carrying capacities of the habitats for each of two mountain ranges in North Carolina*[a]

	Great Smoky Mountains (altitudinal overlap narrow)	Balsam Mountains (altitudinal overlap wide)
$\alpha_{g,j}$	0.194	0.142
$\alpha_{j,g}$	2.255	0.629
K_j	45.27	47.43
K_g	11.69	10.93

[a]The carrying capacities are in terms of the number per plot search, which is known, rather than the total number present.
Source: Hairston (1980a).

but it came only at the end of the fourth year of the experiment (Fig. 5.3). There was also an increase in the proportion of young *P. jordani* on the plots from which *P. glutinosus* was removed, but the increase was to a significantly lower level than in the Great Smoky Mountains. The results confirm the original hypothesis that interspecific competition is much stronger in the region where the altitudinal overlap is narrow than in the region where the overlap is wide. It was somewhat surprising that positive results were obtained in the Balsam Mountains, indicating that competition takes place there, even though it is less intense than in the Great Smoky Mountains.

If we make the assumption that the competition can be described adequately by the Lotka–Volterra equations, it is possible to calculate the coefficients of competition directly. Thus, the reduction of the population of *P. jordani* to a level eighteen individuals below that on the controls during the last three years in the Great Smoky Mountains resulted in an increase in the population of *P. glutinosus* of 3.52 above the controls. The ratio, 0.19, is the competition coefficient, α (see Chapter 1). Once the competition coefficients are known, the Lotka–Volterra equations can be used to calculate the carrying capacities of the two environments for both species (Table 5.1). The similarities of the carrying capacities of the two species in the two mountain ranges are of special interest, because the result indicates that I had been successful in my attempt to select similar habitats in which to carry out the experiments.

The search for niche partitioning and a limiting resource

The experimental results clearly showed interspecific competition, because neither species is a predator on the other (Powders & Tietjen 1974), and predation would be the only way other than competition to account for the mutually negative effect. However, the experiments do not give a clue to the limiting resource for which the two species compete. That was sought in three other studies. The first of these was indirect. Coexisting with *P. glutinosus* and *P. jordani* on the plots were four other terrestrial salamander species in the Balsam Mountains and five others in the Great Smoky Mountains. All of these species prey on the same array of food items: insects and other small invertebrates of the forest floor and low vegetation. In neither mountain range did the removal of either of the large plethodons have an effect on the remaining salamander species, either alone or in combination (Hairston 1981). There was thus indirect evidence that food is not the limiting resource for which *P. glutinosus* and *P. jordani* compete.

Direct evidence was provided from a study by my former student, Dr. Sarah Stenhouse. She investigated the possibility that the lower level of competition in the Balsam Mountains was due to partitioning of the food resource (Hairston, Nishikawa, & Stenhouse 1986). Her method was to capture specimens during their nighttime foraging and shortly thereafter have them regurgitate any food in their stomachs by forcing water into them through a fine plastic tube (Fraser 1976a). There was considerable overlap in kinds of food eaten, although given the array of food types, the distribution was significantly different between the two species of salamanders. The most important part of the findings was the fact that there was a greater similarity of food types in the Balsam Mountains than in the Great Smoky Mountains. For one of the many overlap indices (Southwood 1978) the values were 0.721 for the two species in the Balsams and 0.442 in the Smokies, when the possible range is from zero for no food in common to 1.0 when the diets are identical. Although a probability estimate cannot be assigned to the difference, it is exactly the reverse of expectations based on food partitioning in the Balsams. This was the first time that the diets of coexisting salamanders had been compared in a way that used an a priori test of the meaning of the findings. Previous studies, including mine on the genus *Desmognathus*, had relied on either intuitive interpretations or a posteriori explanations of the observations (Hairston 1949; Dumas 1956; Whitaker & Rubin 1971; Powders & Tietjen 1974). More recent and elaborate examples are the studies of Krzysik (1979) and Lynch (1985), who used the overlap indices as coefficients of competition without independent demonstrations that competition was occurring.

Stenhouse also compared foraging microhabitats in the two mountain ranges. Foraging takes place mostly on the forest floor, but both species can be found climbing herbs and low shrubs on especially humid nights. In neither the Great Smoky Mountains nor the Balsam Mountains did *P. glutinosus* and *P. jordani* differ in the proportion of individuals climbing. On both mountains, the species differed significantly in the proportion on herbs versus woody vegetation. The larger *P. glutinosus* was found more frequently on shrubs and tree trunks, which provide more support for its heavier body.

If climbing were related to partitioning of the foraging microhabitat feeding niche, one would expect the species to be more different in the narrow altitudinal zone where they occur together in the Great Smoky Mountains than they are above the zone (*P. jordani*) or below it (*P. glutinosus*). Stenhouse investigated this possibility. Within the zone of overlap, 19 of 132 *P. jordani* (14 percent) and 16 of 148 *P. glutinosus* (11 percent) were observed climbing. Above the overlap, 11 of 66 *P. jordani* (17 percent) were climbing, and χ^2 for the difference is 0.18 for one degree of freedom. The probability of a greater value by chance is between 0.5 and 0.7, and it is unlikely that a statistical type II error is involved in accepting the null hypothesis that the proportions are not different. For *P. glutinosus* immediately below the altitudinal overlap, none of 109 were climbing – highly significantly different from the proportion in the overlap ($\chi^2 = 12.57$, d.f. $= 1$, $p < .005$). This difference is the opposite of that predicted from niche partitioning and is assumed to be due to some unrelated factor, as there were ample trees and shrubs below the overlap zone.

An interesting aspect of her work is that Stenhouse expected to find evidence of resource partitioning – a reasonable expectation, based on the dominant theory of community organization. As is shown in Chapter 7, her results were consistent with the subsequent outcome of experimental manipulations testing for the evolution of competitive ability. With food and foraging microhabitat eliminated, the resource for which the two species compete was still unidentified. It was suggested from both negative and positive evidence.

The requirements that are likely to be important for terrestrial salamanders are food, refuges, and nesting sites. As far as observations of the natural history are concerned, the species of the forest floor share not only food, as was described earlier, but also the burrows that they use as daytime refuges. Detailed analysis of burrow use has not been made with the care taken in the observations on food and foraging microhabitats, and it is possible that individual burrows are somehow different between the species, but there is no evidence supporting such a speculation. If burrows prove to be the limiting resource, the kind used by *P. glutinosus* and *P. jordani* would have to be different from those used by the remaining four or five species of the forest floor, because removal of either of those two species of *Plethodon* has no effect on the remaining members of the community (Hairston 1981). Nevertheless, evidence is accumulating that in some way space is limiting.

Defense of space

Many of the elements of the behavior of salamanders can be observed conveniently in the laboratory (Noble & Brady 1930; Organ 1958, 1960; Arnold 1976; Jaeger, Nishikawa, & Barnard 1983). Nishikawa (1985a,b) has carried out a series of experiments to assess the importance of intraspecific and interspecific aggression by specimens of *P. glutinosus* and *P. jordani* from both the Great Smoky Mountains and the Balsam Mountains. Like their smaller congener *P. cinereus* (McGavin 1978; Jaeger & Gergits 1979), the large plethodons mark their substrates with skin secretions and feces and after doing so will defend the marked area. Nishikawa established the fact that when intruders introduced into the experimental chambers were of the same sex as the resident, the latter was aggressive toward them. The aggression was observed to be both toward individuals of the same species and those of the congeneric species. In the field, marked individuals of both species were faithful to small areas on the forest floor, some of them over two full seasons of activity (Nishikawa 1985b). All of these behavioral observations are consistent with territoriality, and only observations of active defense of the home range are lacking to establish it firmly. These experiments are relevant to the evolution of both species under the influence of interspecific competition, and they are described in detail in Chapter 7. Their importance here is the demonstration that space is the most likely resource for which these salamanders compete.

From the initial discovery that *P. glutinosus* and *P. jordani* occupy altitudinal distributions that are nearly mutually exclusive, the explanation has been pursued through the hypothesis of interspecific competition and its variable nature on different mountain ranges. Experimental tests confirmed the hypothesis, but raised the question of the nature of the resource for which they compete, as well as the evolutionary route by which they have arrived at their present status in the two areas of differing intensities of competition. Further detailed studies and experiments have established that food is not the object of competition, nor is foraging microhabitat. The difference in intensity of competition was found to be related to relative aggression in the two areas. Combined with their fidelity to small areas on the forest floor, the combination of observations and experiments shows that competition is for space. The critical nature of the space is not yet established, but I have suggested that it may be that nest sites are being defended.

Small plethodons: asymmetric competition

Field observations

When Highton and Worthington (1967) described *Plethodon shenandoah*, they pointed out that it was known only from three mountains in Virginia and, on those, only from north-facing talus slopes. This very restricted ecological distribution attracted the attention of Jaeger (1970), who carefully mapped the exact locations of 315 specimens of *P. shenandoah*, as well as 507 of the related *P. cinereus* – a species widely distributed in eastern North America (Fig. 5.4). With few exceptions, the two species occupy mutually exclusive distributions, even in minute detail. Jaeger could account for the exceptions. The *P. shenandoah* outside the talus were all individuals that were larger than *P. cinereus* adults, and they had eaten prey that were larger than the prey of *P. cinereus*. The *P. cinereus* within the boundary of the talus had evidently spread via the packed soil introduced in the construction of a trail. The explanation offered for the distribution was that the talus was an unsuitable habitat for *P. cinereus*, but that *P. shenandoah* was able to tolerate the parts of the talus slope where at least some soil had accumulated. The hypothesis was proposed that it was unable to invade the area of deep soil because *P. cinereus* was a superior competitor. Jaeger pointed out that in a few places completely surrounded by talus, soil had accumulated to a sufficient depth to remain moist throughout a normal summer – an average of 13.9 cm, as compared with 0.5 cm over most of the talus slope and 46.3 cm for the area outside the talus. In such restricted areas, *P. shenandoah* was found in abundance, suggesting that the soil itself was acceptable as a habitat.

Experiments

The hypothesis was tested in the first experiments on interspecific competition among salamanders (Jaeger 1971a,b). Sets of 1-mm screen enclosures were sunk into each of the four substrates: type I talus (completely free of soil); type II talus (with very shallow soil); type III talus (with deeper soil, as described earlier); and soil outside the talus. In each substrate, three enclosures were used. Two of these were 1 m long by 50 cm wide by 50 cm deep, and each received ten specimens of one of the two species; the third was

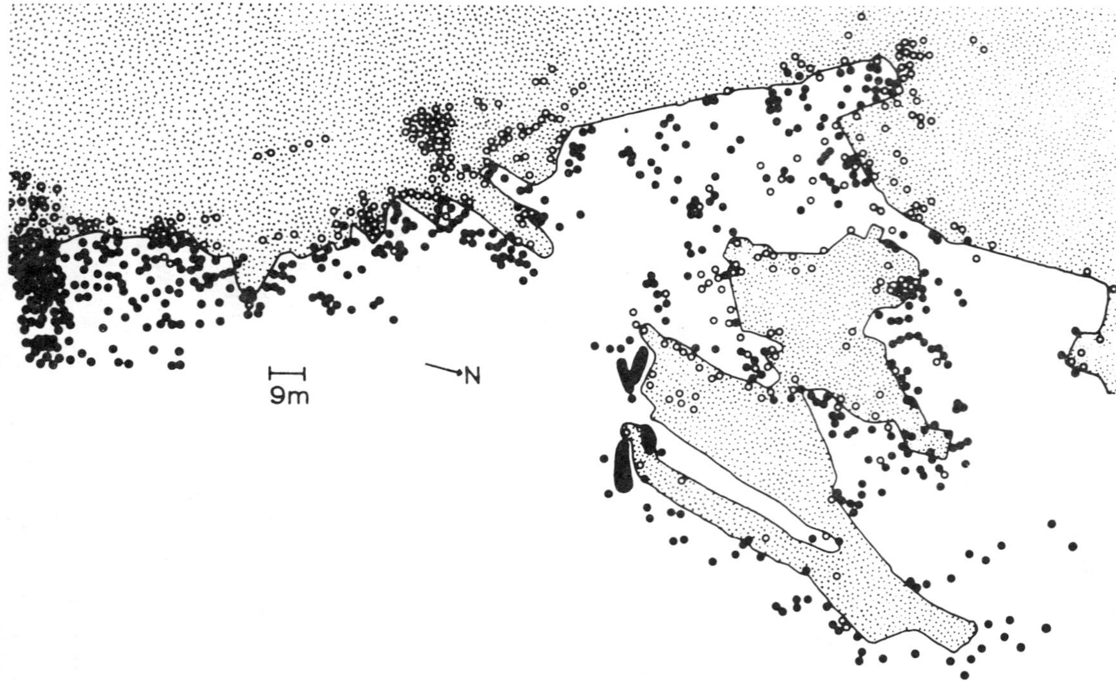

Figure 5.4. A portion of the talus–soil interface on Hawksbill Mountain. Solid lines indicate the interface, stippled areas are talus, and the white area is deep soil. Open circles (*P. shenandoah*) and filled circles (*P. cinereus*) represent the positions of all collected salamanders in this area. Large dark objects are large boulders. (From Jaeger 1970.)

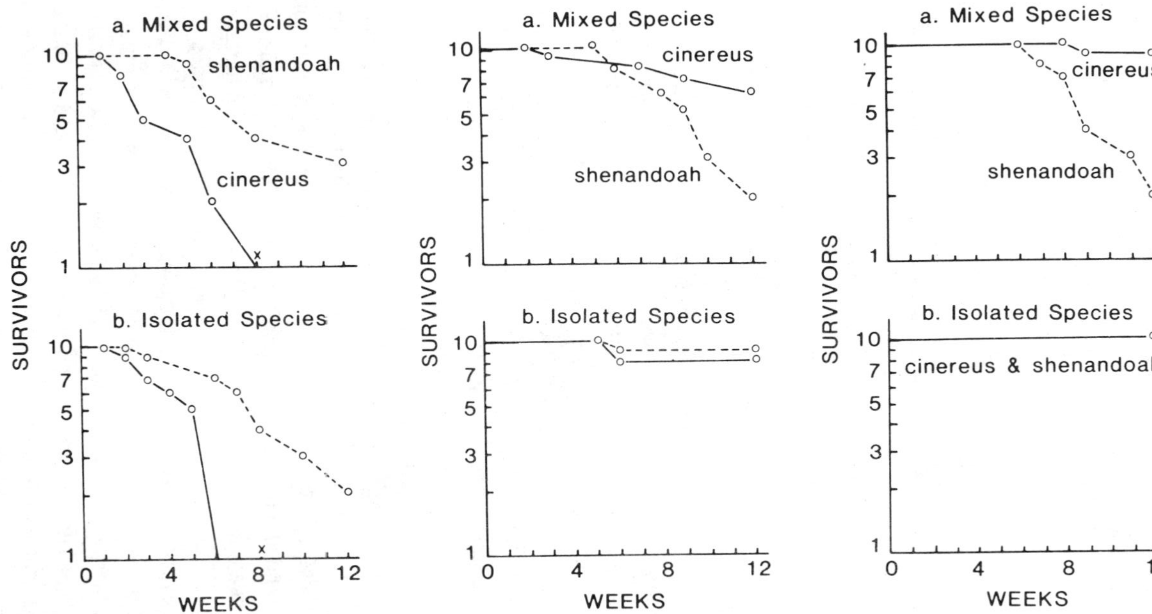

Figure 5.5. Survivorship curves for mixed (top) and isolated (bottom) populations of *P. cinereus* (solid lines) and *P. shenandoah* (broken lines). Pairs of graphs: left, type II talus; middle, type III talus; right, soil. (From Jaeger 1971a.)

Table 5.2. *Results of one-tailed Kolmogorov–Smirnov two-sample tests comparing survivorship of* P. cinereus *and* P. shenandoah *in enclosure experiments*

Habitat	K_D values comparing P.c. isolated/ P.c. mixed	P.s. isolated/ P.s. mixed	P.c. isolated/ P.s. isolated	P.c. mixed/ P.s. mixed
Type II talus	2 ($p > .2$)	1 ($p > .2$)	6 ($p = .05$)	5 ($p = .1$)
Type III talus	2 ($p > .2$)	7 ($p = .01$)	1 ($p > .2$)	4 ($p > .2$)
Soil	1 ($p > .2$)	8 ($p < .01$)	0 ($p > .2$)	7 ($p = .01$)

Source: Jaeger (1971a).

1 m wide and received ten of each species, thus preserving a constant density for all conditions. Each enclosure was examined weekly for survival of the salamanders. None survived the first week in type I talus; the results in the remaining three substrate types are shown in Figure 5.5. *P. shenandoah* survived as well as or better than *P. cinereus* in all substrates when alone; in type II talus, its survival was better, with marginal significance. In enclosures where both species were present, *P. cinereus* had significantly superior survival in soil, better but not significantly so in type III talus, and worse but not significantly so in type II talus (Table 5.2).

The superiority of *P. shenandoah* when alone on talus suggested that its ability to retain water under dry conditions was the basis of its ability to avoid extinction because of the competitive superiority of *P. cinereus* in soil. Jaeger (1971b) suspended ten single specimens of each species in 1-mm-mesh screen containers in a refrigerator at 17 °C and 20 percent relative humidity, with air circulation of 0.2 m/sec. Ten control specimens of each species were under identical conditions, except that the bottoms of the containers were slightly submerged in water. All specimens had been in saturated atmospheres for seven days before the experiment. They were each weighed hourly after the start of the experiment, with the results shown in Figure 5.6. In the dry air, both species lost weight at approximately constant but different exponential rates. At the end of nearly all time intervals, *P. cinereus* had lost a significantly greater percentage of its initial body weight.

The search for the limited resource

Jaeger's experiments were convincing in establishing that the field distribution is brought about by the competitive superiority of *P. cinereus* on

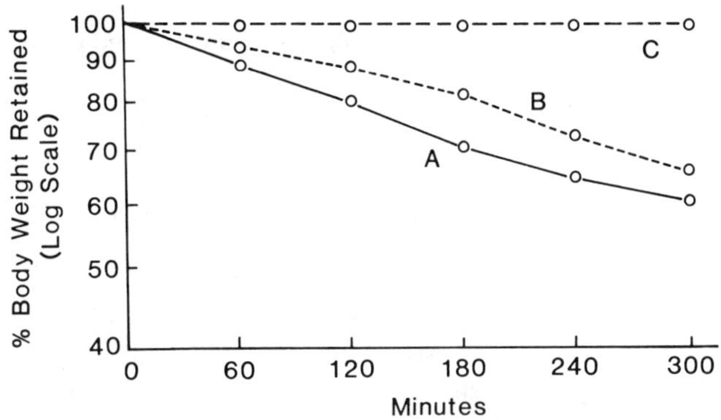

Figure 5.6. Dehydration rates of *P. cinereus* (A), *P. shenandoah* experimentals (B), and controls for both species (C), expressed as percentage body weight retained by survivors, measured per hour. Log scale on vertical axis. (From Jaeger 1971b.)

substrates where it is physiologically capable of surviving and by the physiological superiority of *P. shenandoah* under dry conditions in the talus. Further experiments were required before the mechanism of the greater competitive ability of *P. cinereus* was established. Jaeger (1972, 1974) conducted several experiments testing the hypotheses that the competition is for food or that mutual avoidance is the mechanism. The results were inconclusive. In laboratory comparisons of the proportion of experimental prey (termites) eaten by each species, no significant difference was found, either when they were required to forage separately or together, and in tests for differential distribution in experimental terraria, *P. cinereus*, the evident winner in the field, shifted its position in the presence of *P. shenandoah*, but there was no effect on the latter. In a more elaborate set of experiments, Kaplan (1977) tested the prey-catching ability of *P. cinereus* and *P. shenandoah* separately and found that *P. cinereus* consumed *Drosophila melanogaster* at a faster rate than did *P. shenandoah*, whether in a simple or a physically complex environment. When the two species were tested together, however, *P. shenandoah* increased its feeding rate to the point where it captured significantly more flies per unit of time than did *P. cinereus*. The experiment gave the opposite result from that expected on the basis of Jaeger's field experience.

At about the same time, Fraser (1976a), working with *P. cinereus* and *P. hoffmani*, a species closely related to *P. shenandoah*, found that during dry weather, food was unavailable to either species, as they were confined under cover objects. However, he was unable to show significant interspecific effects in laboratory experiments, and he suggested that competition for space was as viable an explanation of the observed field distributions as was competition for food. The temporally varying availability of food for these species continued to be explored (Jaeger 1978, 1979), but evidence was accumulating that behavior activated by chemical secretions and the chemical sense is important to this group of salamanders.

Aggressive behavior of P. cinereus

Grant (1955) had observed aggressive defense of small areas in aquaria by the plethodontids *Eurycea bislineata* and *Hemidactylium scutatum*, and Thurow (1976) made similar observations on both large and small plethodons. Madison (1975) found that *P. jordani* could distinguish among the odors of conspecific individuals, depending on proximity of residence. The association between chemical marking by resident specimens of *P. cinereus* and the use of the nasolabial grooves to detect the marked substrate followed (Tristram 1977; Jaeger & Gergits 1979). Jaeger and Gergits further showed that both *P. cinereus* and *P. shenandoah* could distinguish each other's secretions and preferentially chose the substrate marked by themselves.

Territoriality in P. cinereus

When individuals of each species were placed together under conditions of food limitation, aggressive behavior could be observed (Wrobel, Gergits, & Jaeger 1980). The amount of food provided was set at 75 percent of what had been found necessary to maintain body weight. An individual was defined as the "winner" if it gained more or lost less weight than the matched specimen. *P. cinereus* winners initiated aggressive behavior significantly more frequently than did losers, whether with conspecifics or with *P. shenandoah*, and they also performed significantly more frequent aggressive acts than did *P. shenandoah* when the two species were together, regardless of weight gain.

Much research on *P. cinereus* has been directed toward the basis of individual recognition by means of pheromones. An individual placed in a container on a moist substrate will begin to explore the area and to tap the substrate with its nose, thus touching the nasolabial grooves to the ground. These are narrow grooves connecting each of the nostrils with the mouth across the upper lip. They frequently end in protrusions of the lip. The function is evidently to carry chemical information about the substrate to the nostril. This nose-tapping behavior is especially prominent when a specimen is placed in a container that has been occupied by another individual for several days (Jaeger & Gergits 1979). The information is evidently chemical, because surrogate "salamanders" made of rolled-up absorbent paper and impregnated with secretions from the floor of another salamander's container could alter the behavior of the resident individual (Jaeger et al. 1983), but a living intruder elicited a stronger response. The information carried by the pheromones includes not only species and sex (Jaeger & Gergits 1979) but also whether the individual had occupied a nearby territory or was a stranger (Jaeger 1981).

This last information came from Jaeger's demonstration that individual *P. cinereus* behave less aggressively toward individuals to whose secretions they have been exposed repeatedly than toward those whose secretions are unfamiliar. The experimental design was as follows: Chambers 18 × 13 × 8.5 cm were provided with soil topped by wet paper towels and divided into three linearly arranged compartments that could be isolated or not. One salamander was placed in each end compartment, and on alternate days each was given sole access to the central compartment. They were thus able to become familiar with each other's secretions, but were not allowed direct contact. After six days, two sets of experiments were carried out, one set in which pairs of salamanders were placed in arenas the floors of which had fresh paper towels (no prior ownership), and another set in which one of the salamanders had resided for seven days. For each set of conditions, salamanders were paired with "familiar" and "unfamiliar" individuals.

When neither specimen had established ownership of the arena, thirty-minute encounters were observed, with the behavior of both individuals being recorded as threatening or submissive postures or as attacking (biting). Threatening posture involved raising the whole trunk above the substrate; submissive posture involved flattening the trunk on the substrate. A noninteracting salamander normally has only the head and fore part of the body raised. Under these conditions of nonownership, the 164 salamanders observed behaved in significantly different ways toward familiar and unfamiliar individuals. Threat was displayed more toward the unfamiliar individuals by 65 percent of the specimens, with 32 percent doing the opposite, and 3 percent showing no difference. Submissive display was almost exactly the reverse: 61 percent spent more time in submissive displays toward familiar salamanders than toward unfamiliar ones, and only 37 percent showed the opposite. Biting was uncommon, but of the nineteen salamanders that bit during these encounters, thirteen bit unfamiliar individuals a total of thirty-four times, whereas five administered eleven bites to familiar ones. One specimen bit one of each kind (Fig. 5.7).

While one of the individuals was establishing ownership of the arena, the

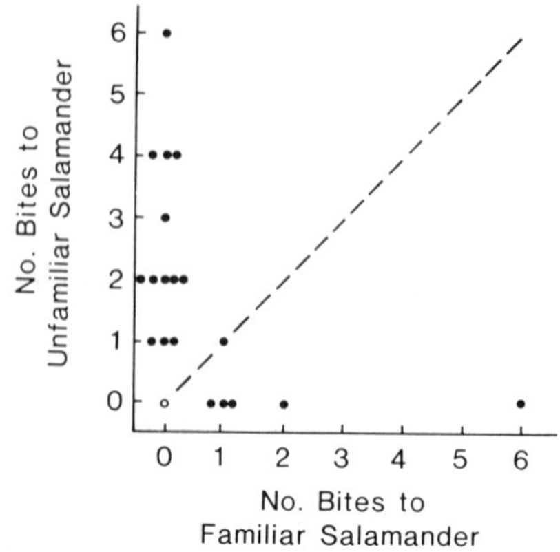

Figure 5.7. Each filled circle represents the number of bites administered by an individual salamander to an unfamiliar conspecific in one encounter (ordinate) and to a familiar conspecific in another encounter (abscissa) when neither paired salamander had prior ownership of the area. Points are slightly offset horizontally when more than one point falls at the same locus. The open circle represents 145 salamanders that did not bite in either encounter; the dash line denotes hypothetically equal responses to familiars and unfamiliars. (From Jaeger 1981.)

previously described surrogate that had been rubbed over the substrate conditioned by one other specimen was introduced each day into its arena, thus making one other individual "familiar." The results of the experiments after a second salamander was introduced into these arenas were dramatic; only biting was recorded, and for only five minutes. In contrast to the experiments where no ownership had been established, fourteen of twenty-three owners bit the intruders, and delivered a significantly larger number of bites to unfamiliar ones (mean = 4.9) than to familiar ones (mean = 1.5) (Fig. 5.8). Most of the attacks were bites to the snout or to the tail (Fig. 5.9). A bite on the tail frequently resulted in autotomy, thereby costing the victim fat reserves; a bite to the nasolabial grooves damages the victim's ability to sense prey. In a feeding experiment, twelve "losers" that had been bitten on the snout ate an average of only 5.8 freshly killed *Drosophila* in two hours; twelve "winners" ate an average of 18.6; and six tail-bitten "losers" ate an average of 20.2. Field-collected specimens that had scars over the snout weighed significantly less per unit length than did those without scars.

These experiments established "dear enemy" behavior in *P. cinereus*, the interpretation of which is that familiar salamanders that have established mutual territorial boundaries save energy and avoid risk of injury by "respecting" those boundaries. The territoriality of *P. cinereus* seems well established, and other studies by Jaeger and his colleagues have clarified the use of

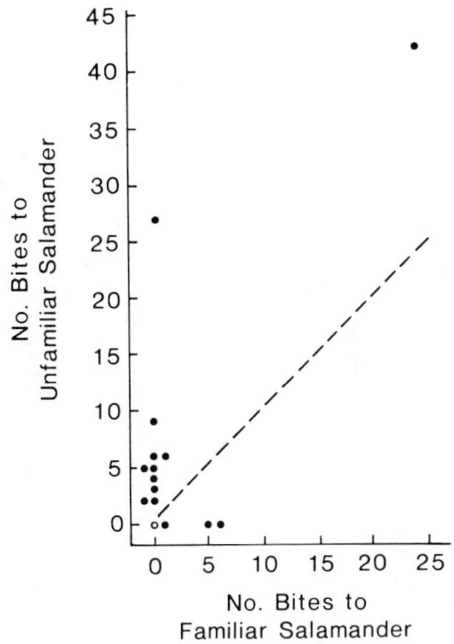

Figure 5.8. Numbers of bites given by each of fourteen salamanders (filled circles) to unfamiliar (ordinate) and familiar (abscissa) conspecifics when the biting salamander had prior ownership of the area. Open circle represents nine salamanders that did not bite in either encounter. (From Jaeger 1981.)

territories in this species. Following rains, the salamanders forage in the litter on the forest floor, where prey are abundant. Dry weather has relatively little effect on the abundance of the prey, but restricts the ability of the wet-skinned salamanders to forage far from refuges under logs and stones. Shortly after emerging from underground winter retreats, several specimens can be found together under a single log or stone. Later, these refuges tend to be occupied by solitary animals (Jaeger 1979). The availability of prey during dry weather is restricted to these damp locations, and it is likely that the marking and defense of the refuges constitute the basis of territoriality. From the standpoint of the interspecific interactions that are the mechanisms by which communities are determined, the interspecific territoriality shown by *P. cinereus* and *P. shenandoah* is an excellent example of the effect of interspecific competition for the refuges.

Noncompetitive relationships among terrestrial salamanders

The two examples in which interspecific competition has been demonstrated by field and laboratory experiments should not obscure the fact that among terrestrial salamanders there have been more studies in which it was

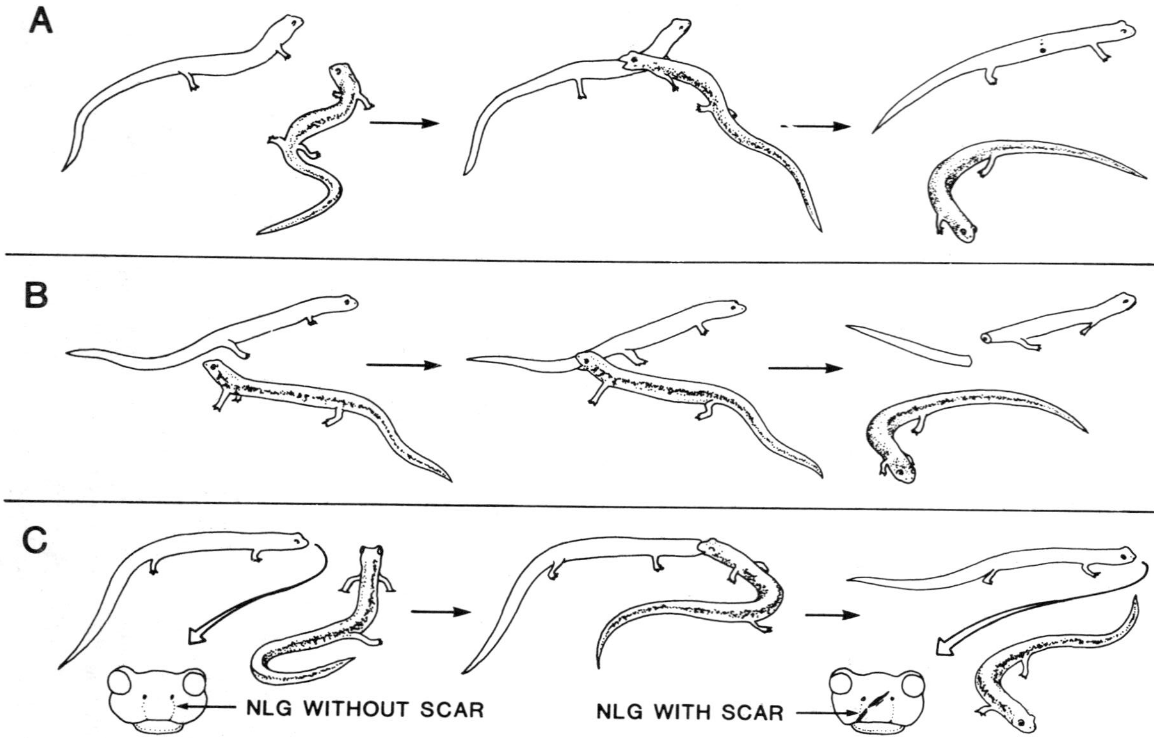

Figure 5.9. Areas of body attacked during aggressive encounters: A, trunk; B, tail; C, nasolabial grooves (NLG) on the snout. (From Jaeger 1981.)

concluded that the species were not competing. My own example was mentioned earlier (Hairston 1981), but it was preceded by several others.

The first of these was by Dumas (1956), who worked with the western *Plethodon dunni* and *P. vehiculum*. The two species occur together on talus slopes in a wide area in west-central and northwestern Oregon. *P. vehiculum*, the smaller, has a distribution extending northward from the area of overlap through western Washington State. *P. dunni*'s range includes southwestern Oregon, from which its congener is absent. Extensive collecting in the common range showed slight differences in location with respect to substrate moisture or temperature where individuals were found, with *P. dunni* found more frequently on "wet" substrates, and *P. vehiculum* in locations that were more variable in both temperature and relative humidity. In laboratory experiments, *P. vehiculum* was more tolerant to high temperatures and low humidities, but more sensitive to low temperatures – a surprising result, considering the relative sizes and distributions of the two species. *P. dunni* had eaten a greater variety of prey items, but there was much overlap. A variety of other features of behavior and local distribution were recorded, without any significant differences being observed. No competition experiments were performed, but it was found that in a device where two individuals were confined in a trough in which the humidity could be controlled, the two species chose the same humidity when alone. When they were present together, *P. dunni* caused *P. vehiculum* to select a humidity lower than its preferred one. The overall conclusion was that competition between the species was slight and that adverse physical conditions kept the populations below levels at which they would compete. This conclusion seems to be intuitive, rather than being based on satisfactory evidence.

Fraser (1976b) conducted a study rather similar to that of Dumas (1956). Two species of *Plethodon* of distinctly different adult sizes, *P. hoffmani* and *P. punctatus*, coexist in the mountains of Virginia and West Virginia. Fraser compared their use of potentially limiting resources: food, cover objects, foraging times, and activity seasons. Although the adults of the two species were sufficiently different in head width to meet the size-difference criterion of Hutchinson (1959), and in fact overlapped in food utilization by a sufficiently small amount to meet the mathematically derived criterion of May and MacArthur (1972), the young of the larger *P. punctatus* overlapped with all sizes of *P. hoffmani*, both in head width and in food utilization. Fraser proposed that there was sufficiently different distribution of the underground retreats of the two species that the actual times of foraging differed enough to preclude competition for food. How the prey populations could fail to be affected by both species of predators, even though the latter foraged at different times, was not explained. As with the study by Dumas (1956), no experimental test for the existence of competition was performed, and the conclusion that competition was minimal rested on the criterion of apparently sufficient niche partitioning.

Communities of terrestrial genera other than *Plethodon*

Two conspicuous faunas of terrestrial salamanders are found outside eastern North America. They are the Tribe Bolitoglossini in the neotropics and a mixture of the plethodontine genera *Aneides* and *Ensatina* and the bolitoglossine genus *Batrachoseps* in the western United States. It is true that several species of *Plethodon* occur in western North America, and one species of *Aneides* occurs in the eastern United States, but the community relationships of these have not been studied intensively, except for the work of Dumas (1956) reported earlier.

Neotropical Plethodontidae

Despite the general agreement that salamanders are unique among poikilothermic terrestrial vertebrates in being characteristic of the north temperate zone and that they almost certainly arose and diversified there, approximately half of all salamander species are found in the area from Mexico to Brazil. The vast majority of these are representatives of a single radiation from one of the two purely terrestrial groups of the Family Plethodontidae. The comparative ecology of these 150 or more species becomes less daunting when it is realized that most of them have restricted geographic distributions, and within a geographically defined area there is much altitudinal segregation. There have been two attempts to carry out sufficiently detailed studies of local distributions to begin to define the problems in community ecology. The studies were by Schmidt (1936) and Wake and Lynch (1976).

It was Schmidt's use of a single vertical transect on the volcano Tajumulco in Guatemala that was the inspiration for my own early work in the southern Appalachians. His transect extended from sea level to the top of the mountain at 4,220 m, and Schmidt recorded relative abundances of each of nine species at successive elevations (Fig. 5.10). The most striking distributional feature was the sharply defined association of salamander species with vegetational zones, suggesting that factors external to the salamanders were responsible for most of the altitudinal zonation. In the case of some of the species pairs not separated by vegetation, local habitats were decidedly different. For example, "*Oedipus*" *flavimembris* lives on the forest floor, whereas "*O.*" *engelhardti*

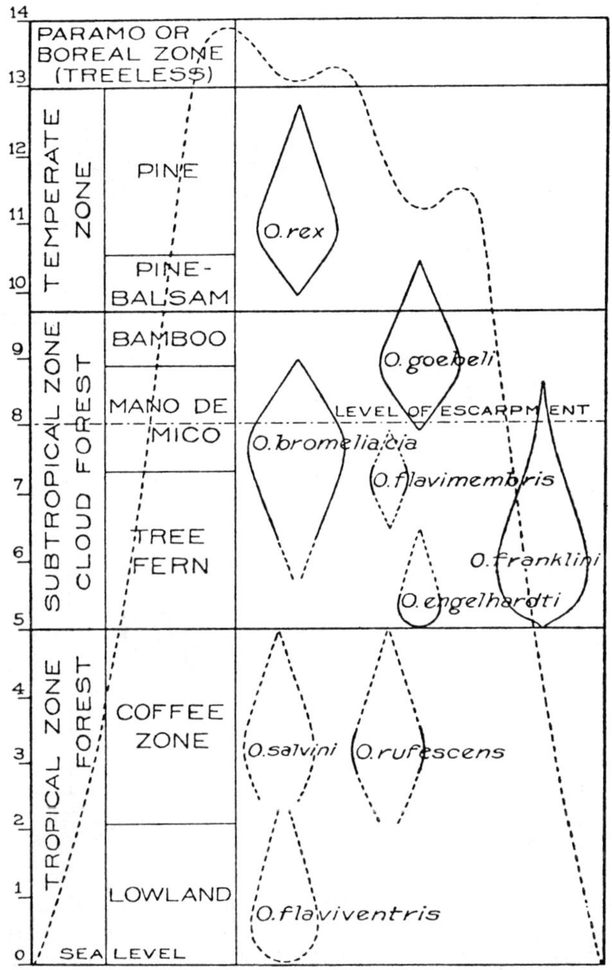

Figure 5.10. Vertical distribution of the species of bolitoglossine salamanders on the volcano Tajumulco, with vegetation zones. (From Schmidt 1936.)

is an inhabitant of the leaf bases of the epiphytic bromeliads. At the time when Schmidt wrote, all tropical plethodontids were placed in the genus *Oedipus*, a name that has since been found to be preoccupied (Taylor 1944). Schmidt made no attempt to relate altitudinal distributions to mutual ecological interactions, but in the light of subsequent work on salamander distributions, it is tempting to suggest that some of the inversely changing abundances may be attributable to interspecific competition.

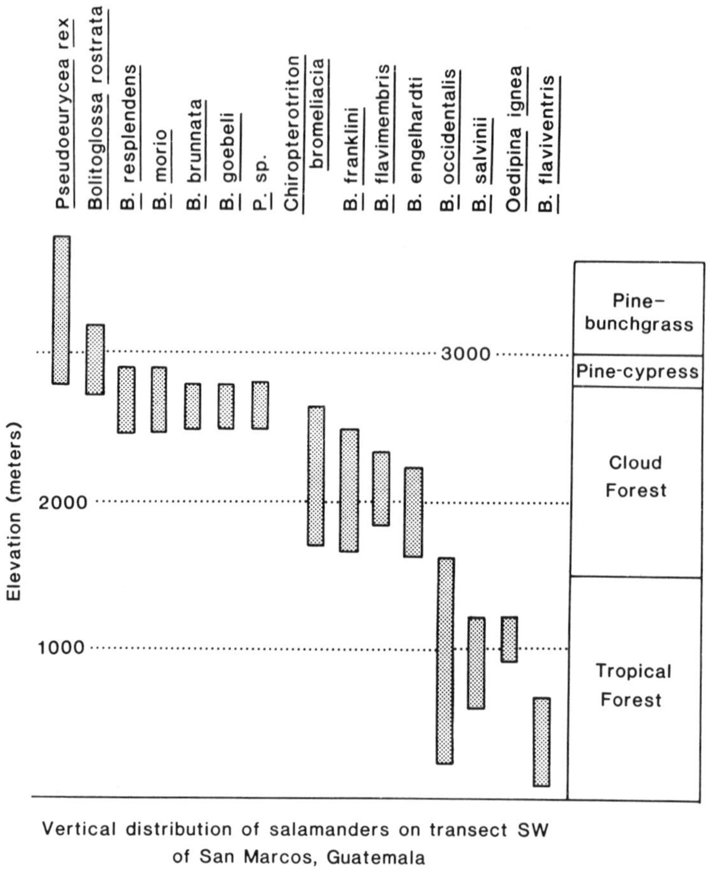

Figure 5.11. Altitudinal distribution of fifteen species of plethodontid salamanders along a transect near San Marcos, Guatemala. (Adapted from Wake & Lynch 1976, with some altitudinal ranges extended to include data provided by D. B. Wake.)

Forty years after the appearance of Schmidt's study, Wake and Lynch (1976) published on much more extensive surveys throughout the distribution of the tropical plethodontids. Their most complete analysis of local distributions was of a transect on the same volcano studied by Schmidt and only 15 km distant from his transect. They recorded fifteen species (Fig. 5.11), including all of those found by Schmidt (*Bolitoglossa occidentalis* in their data represents a splitting of *B. rufescens*, the name used in the older paper). The two transects agree well on the detailed altitudinal distributions of the species recorded in both studies. Of the six species not recorded by Schmidt, two represent range extensions from the Guatemalan plateau, where Schmidt found them; they are *B. rostrata* and *B. morio*. *B. resplendens* is a range extension from Mexican localities not explored by Schmidt. Two species of *Pseudoeurycea* (*P. brunnata* and a species undescribed in 1976) were unrecognized by Schmidt. Wake and Lynch also added *Oedipina ignea*, a rare fossorial species. As in the earlier study, vegetational and microhabitat segregations were prominent, leaving as potential coexisting competitors the following combinations: *P. brunnata* and *P. goebeli* on the ground in the cloud forest, and *Chiropterotriton bromeliacea*, *Bolitoglossa franklini*, and *B. engelhardti* as potential arborial competitors in the same forest. Those in the latter group have distinctly different sizes. Wake and Lynch concluded that competition among sympatric forms was unlikely. Altitudinally, there were some parapatric distributions that were reminiscent of *Plethodon glutinosus* and *P. jordani*, as well as of *P. cinereus* and *P. shenandoah* in the Appalachians. *Bolitoglossa occidentalis* and *B. engelhardti* have altitudinal distributions that are nearly in contact at the lower edge of the cloud forest, as do *B. franklini* (lower) and the two species of *Pseudoeurycea* (higher) within the cloud forest. These situations would be worth experimental manipulations, as all are common, at least locally.

California communities

Two studies on plethodontid genera in California have concentrated on food as a potentially limiting resource. Maiorana (1978) examined *Aneides lugubris* and *Batrachoseps attenuatus* in four habitats, including one on an island in San Francisco Bay. As with Fraser's (1976a,b) study of *Plethodon punctatus* and *P. hoffmani*, the sizes of adults were very different, and the larger *Aneides* had consumed some prey items well beyond the range of sizes taken by *Batrachoseps*. However, on the island, large prey were relatively unavailable, and the diets of the two species had overlap values in excess of that theoretically tolerable for coexistence on a limiting resource (May & MacArthur 1972). In other habitats on the mainland, Maiorana observed that *Aneides* occupied larger refuges than did *Batrachoseps*, and she concluded that space was more important than food in regulating the populations of these two species. Food influenced both the size of animals on the island and the amount of fat in *B. attenuatus*, as judged by the diameter of the tail – the location of fat storage. She argued that this was evidence of competition for food, but that such competition could result in reduction in body size and

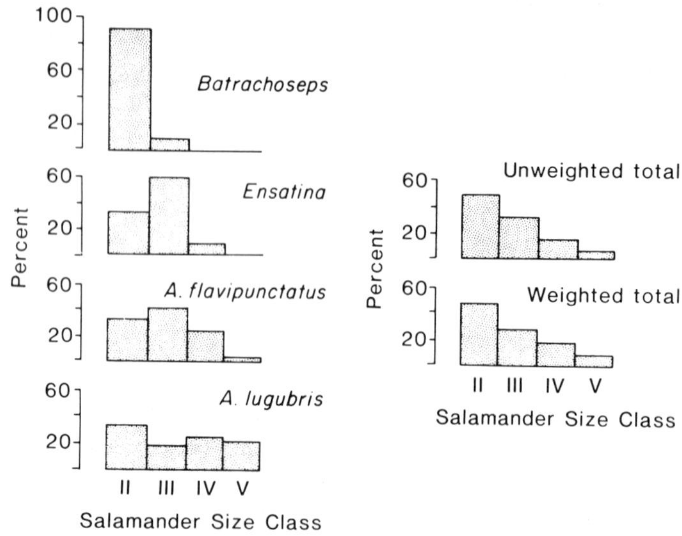

Figure 5.12. Size-frequency distributions of four sympatric plethodontids, based on collections from two areas. Histograms show percentages of individuals that fall into each 15-mm size class, starting at 35 mm. Class I juveniles (20–34 mm standard length) are omitted because of insufficient samples. The unweighted total histogram shows the overall size distribution of salamanders, assuming that the four species are equally abundant. The weighted total histogram shows the distribution of sizes that results if each species is weighted according to its observed abundance in collections. (From Lynch 1985.)

fecundity without resulting in limiting density, because populations of both species on the island had remained high and stable for fifteen years. Such an allocation of the food resource requires a better explanation than was offered.

The most complete analysis of the diets of coexisting salamander species is that of Lynch (1985), who studied four species at one location in northwestern California, three of which were also studied at two other locations. *Aneides lugubris*, *A. flavipunctatus*, and *Batrachoseps attenuatus* were abundant enough at all locations to give meaningful data. *Ensatina eschscholtzii* yielded sufficient data at only one place. The four species were different in size, but overlapped widely (Fig. 5.12). Lynch measured the dimensions of all prey items in the stomachs of the salamanders and compared the diets of the size classes for each species. As all observers have found, there was much overlap in diet between size classes within species and between species. Moreover, the differences that were found involved the relationship between the size of the salamander and the size of the largest prey item (Fig. 5.13). As far as

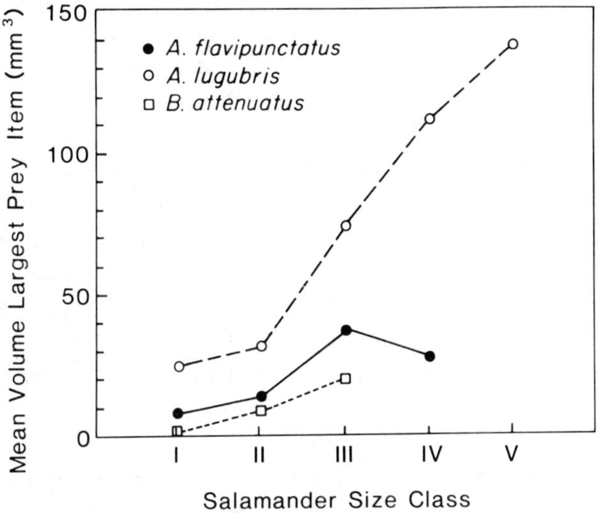

Figure 5.13. Maximum prey volume size versus standard length in three sympatric plethodontids at one location, based on analysis of stomach contents of 109 *A. flavipunctatus*, 110 *A. lugubris*, and 69 *B. attenuatus*. All interspecific differences in maximum prey volume are significant at $p < .05$, except for class I individuals of the two *Aneides* species. (From Lynch 1985.)

competition was concerned, the conclusion was that "direct or indirect competition among coexisting salamanders appears to be a reasonable possibility" (Lynch 1985, p. 347). Given the nature of the study, involving as it did an analysis of overlap of the food niche without any experimental test of competition, the result was almost a foregone conclusion before the study began.

Synthesis

Interspecific interactions among terrestrial salamanders are not found nearly as universally as they are among pond-inhabiting species. In several studies, the evidence can only suggest that we are looking at the evolutionary results of competition that was once intense – "the ghost of competition past," a phrase usually attributed to Connell, but conceded by him (Connell 1985) to have originated with M. Rosenzweig. In such situations, ecological differences are interpreted as representing niche partitioning under the selective influence of interspecific competition – an interpretation that rests on weak inference. We do not know whether the difference is the result of competition or simply the result of evolving in ecologically different places, nor do we know whether or not competition, if any, has been reduced below a level detectable by suitable experiments.

Terrestrial salamanders lack a larval stage, and immature individuals are therefore capable of interacting directly with adults, both conspecific and heterospecific. The obvious size differences may be important if there is competition for food, because large salamanders are capable of consuming some prey that are too large for smaller specimens. Where coexisting species are of significantly different sizes as adults, the fact that several size classes of both species are present may mean that only the adults of the larger species are exempt from competition for food, provided that food is limiting in amount to both species. Unequivocal demonstration that food is limiting has not been provided, but there is evidence that the territoriality of *Plethodon cinereus* may represent a response to periodically occurring environmental conditions. The species cannot forage safely in the leaf litter of its forest habitat during dry periods, and defense of the damp areas underneath cover objects seems to be involved in its territoriality. Such a system could make for behavioral interactions between adults and young, even though food shortage was not directly involved.

Whatever the basis, the two experimentally demonstrated examples of competition among terrestrial salamanders have both involved competition for space. In both examples, competition was first suspected because of the distributional relationships of the two species involved – altitudinal exclusion in one case and narrow parapatry correlated with substrate type in the other. In both, the mechanism has been found to be through aggressive interference. The work of Jaeger and his colleagues on small plethodons has shown not only that biting is used but also that the biting is directed preferentially toward parts of the anatomy (nasolabial grooves and tail) that are energetically very important. Our work on the large plethodons *P. glutinosus* and *P. jordani* has related aggression to the independently measured local intensity of competition. The amount of altitudinal overlap on different mountain ranges was in turn determined by the level of aggression. We do not know the resource being defended aggressively, but the method of detecting the level of competition (reciprocal removal of the two competing species) had no effect on the other four or five terrestrial species present. Inasmuch as these other species have diets that overlap broadly with the competing ones, it seems very unlikely that food is the limiting resource or the object of competition.

Outside of the eastern United States, there are two flourishing faunas of terrestrial salamanders: western United States and tropical America. Research on the community ecology in these two areas has been confined to comparative methods, and, as is to be expected, there are indications that competition may be occurring at present or may have occurred in the past, with the result that the potentially competing forms are now sufficiently different in requirements for competition to be below detectable levels. Whatever the case, any assured knowledge of the relationships must await the laborious process of carrying out field experiments, the duration of which must be at least as great as the duration of the longest life cycle.

6

Communities of streamside salamanders

Introduction

Most species of streamside salamanders are classified as amphibious in the array of life histories presented in Chapter 3 (Table 3.1). Ecologically, they are quite different from the pond-inhabiting forms whose community ecology is described in Chapter 4. Streamside species have a range of life histories. Some, such as *Gyrinophilus porphyriticus*, spend up to four years as larvae. At the opposite extreme, *Desmognathus wrighti* completes the larval stage before hatching. Thus, habitat is more consistent than life history in the community ecology of streamside salamanders. Studies of interspecific interactions have been confined to metamorphosed individuals, but there have been good analyses of the population ecology of stream-living larvae (Bruce 1979, 1980, 1982, 1985), and we should not ignore the possibility that interactions among larvae may be crucial to a complete understanding of the community.

The nature of the habitat, with its gradient of moisture, led early workers to concentrate on the comparative physiology of the salamanders living there. Littleford, Keller, and Phillips (1947) found a marked similarity among *Desmognathus fuscus*, *Eurycea bislineata*, and *Pseudotriton ruber* in the proportional loss of body weight that they could tolerate when subjected to drying conditions. The loss was almost entirely water. As they noted, these three species have very similar ecological distributions. Similar but less sophisticated experiments (Hairston 1947, 1949) revealed that three species of *Desmognathus*, distributed differently along the streambank gradient, showed increasing tolerance to desiccation in accord with their ecological distributions. When placed in closed containers with dry paper towels, metamorphosed specimens of *D. quadramaculatus*, the most aquatic, died first, followed in order by *D. monticola* and finally by *D. ochrophaeus*, the most terrestrial of the three species. The individuals were of approximately the same body length.

From the standpoint of interspecific interactions and community ecology, practically all of the work has been with the plethodontid Subfamily Desmognathinae. Dunn (1926) first pointed out the probable evolutionary sequence, as paralleled by the existing species, from a nearly aquatic streambank ancestor to both terrestrial and completely aquatic modes of life. He discussed the radiation into different ecological niches, but did not propose a basis in natural selection for the progression. His monograph set the stage for ecological work on this community.

Ecological differences among the Desmognathinae

The search for ecological differences among related species of Plethodontidae began almost as soon as their abundance in the southern Appalachians was recognized. Dunn (1917a,b, 1926) reported that *Desmognathus quadramaculatus* and *D. monticola* were characteristic of streams of different sizes, with the former in larger streams. He also recognized the tendency for it to be more aquatic. *D. ochrophaeus* was observed to be more variable in habitat than either of its larger congeners, with a tendency toward terrestrialism. Noble (1927) disagreed with Dunn and reported that adults of *D. quadramaculatus* and *D. monticola* segregated on the basis of stream substrate, with *D. quadramaculatus* in rocky streams and *D. monticola* in streams with mud bottoms. He found no such segregation between immature individuals. Noble noticed an increase in terrestrialism of *D. ochrophaeus* with altitude. He was the first to suggest that interspecific competition might be responsible for part of the habitat differences that he found, arguing that the larger *D. quadramaculatus* could drive *D. monticola* away. Dunn (1928) defended his conclusions, and there were brief comments on the habitats of these species by King (1939), who had previously (1936) described *D. wrighti*, a high-altitude species even more terrestrial than *D. ochrophaeus*.

When I began my research on the ecology of plethodontids in 1940, I concentrated on taking quantitative measurements of the locations in which individuals were found, in the hope that some of the issues could be resolved. The only consistent differences that I found among the five species of Desmognathinae in the Black Mountains of North Carolina were in their distributions with respect to streams (see Table 1.1) and in the change in distribution of *D. ochrophaeus* with altitude (Table 6.1). There is a definite progression from aquatic to terrestrial, with a considerable amount of overlap in the distributions of species adjacent to each other in the series. Stomach contents of the four species of *Desmognathus* were examined and were found to be less different than would be expected from their ecological distributions – *D. quadramaculatus* had taken 65 percent of its food from terrestrial situations. This seemed inexplicable at the time, but I have recently found that at night

Table 6.1. Desmognathus ochrophaeus: *ecological distribution at high and low elevations (percentages)*

			Distance from stream (ft)						
Elevation	Stream	Streambed	1–10	10–20	20–40	40–60	60–100	100–200	>200
≥4,500 ft (222)	8.6	14.0	13.5	13.1	6.8	6.8	2.6	5.3	29.3
<4,500 ft (157)	4.5	19.7	35.0	24.2	12.1	1.9	1.3	1.3	0

[a]Numbers in parentheses refer to total numbers of specimens on which the percentages are based.
Source: Hairston (1949).

the species forages extensively away from streams: During summer months, 65 percent of the metamorphosed individuals were terrestrial. The average distance of adults from water was 1.38 m in July and 1.63 m in August, including those still in streams. All four species were observed to climb shrubs and trees during their nocturnal activity, with the terrestrial *D. wrighti* climbing most, and *D. ochrophaeus* next. My interpretation of these observations was that the differences were related to competition between the species. I could not envision speciation through ecological isolation, because that did not exist. I proposed that geographic isolation had been followed by sufficient ecological divergence for the species to coexist after mutual occupation of the same area.

The study by Organ (1961) added much confirmatory evidence to the conclusion that competition had been the important interaction between the species of Desmognathinae. The original implication that the community was saturated by five coexisting species also appeared to be confirmed by the distribution of a sixth species, *D. fuscus*, in the area around Whitetop Mountain, Virginia, where Organ worked. As in the Black Mountains, *D. monticola* was not found along main streams at high elevations, reaching a maximum elevation of 1,300 m. Below that elevation, *D. fuscus* was rare or totally absent. *D. fuscus* occupied an intermediate position between *D. quadramaculatus* and *D. monticola* on the gradient from aquatic to terrestrial and apparently found it difficult to coexist with both of its larger congeners. Above 1,300 m it was abundant along streambanks also occupied by *D. quadramaculatus*. Below 1,300 m, *D. fuscus* and *D. monticola* were found together in seepage areas that were independent of the large streams. The assumption was that *D. quadramaculatus*, the most aquatic of the trio, was unable to traverse the forest floor to reach the seeps from its normal habitat in the streams. Thus, it was Organ's interpretation that any two of the three species could coexist, but that interspecific competition made it difficult for all three to live together. That neat interpretation cannot hold true everwhere, as there are locations where the three species do coexist (J. R. Bailey, pers. commun.; Southerland 1985, 1986a,b).

Since the earlier studies, there has appeared a dichotomy in the interpretation of the structure of the *Desmognathus* community. Tilley (1968) argued that the tendency toward a terrestrial habitat was probably acquired after speciation into larger and smaller forms and that there was selective pressure on the smaller to become more terrestrial in avoiding predation by the larger aquatic species. In a discussion of tests of community theory, I stated that the *Desmognathus* community was as good an example of evidence for niche partitioning as a result of interspecific competition as was likely to be obtained from observational evidence alone (Hairston 1973). In the same year, Huheey and Brandon (1973), commenting on the much greater density of *D. ochrophaeus* on rockfaces than on streambanks in the forest (see Chapter 2) again suggested that predation was resposible for the difference, because the larger, predatory salamanders were not found above the foot of the rockfaces, except in unusual situations. Means (1975) analyzed in detail the local distribution of *D. fuscus* and *D. auriculatus* in northwestern Florida and concluded that he was observing competitive exclusion between the two species. Krzysik (1979) made a thorough study of niche partitioning in a community including three species of *Desmognathus* in Pennsylvania. He recorded microhabitat distribution, cover type, and a detailed analysis of prey sizes for both adult and immature specimens of *D. monticola*, *D. fuscus*, and *D. ochrophaeus*. Following the theoretical work of MacArthur and Levins (1967), May (1973a,b, 1975), Roughgarden (1974, 1976), Lawlor and Maynard Smith (1976), and others, Krzysik calculated niche widths and niche overlaps for most combinations of immature and adult members of the three species. He concluded that the niche partitioning in the community was the result of coevolution under competition, as visualized in the models of Roughgarden (1976) and Lawlor and Maynard Smith (1976). Although he cited the papers of Tilley (1968) and Huheey and Brandon (1973) in other contexts, Krzysik did not discuss the divergence of opinion about the probable selective pressures in the evolution of the genus.

In thinking about the ecological interactions structuring the *Desmognathus* community, I became convinced that my earlier conclusion that competition was primarily responsible was questionable. The arguments of Tilley (1968) were convincing, because the relative sizes of species along the gradient from aquatic to terrestrial was the reverse from what would be expected of the extra

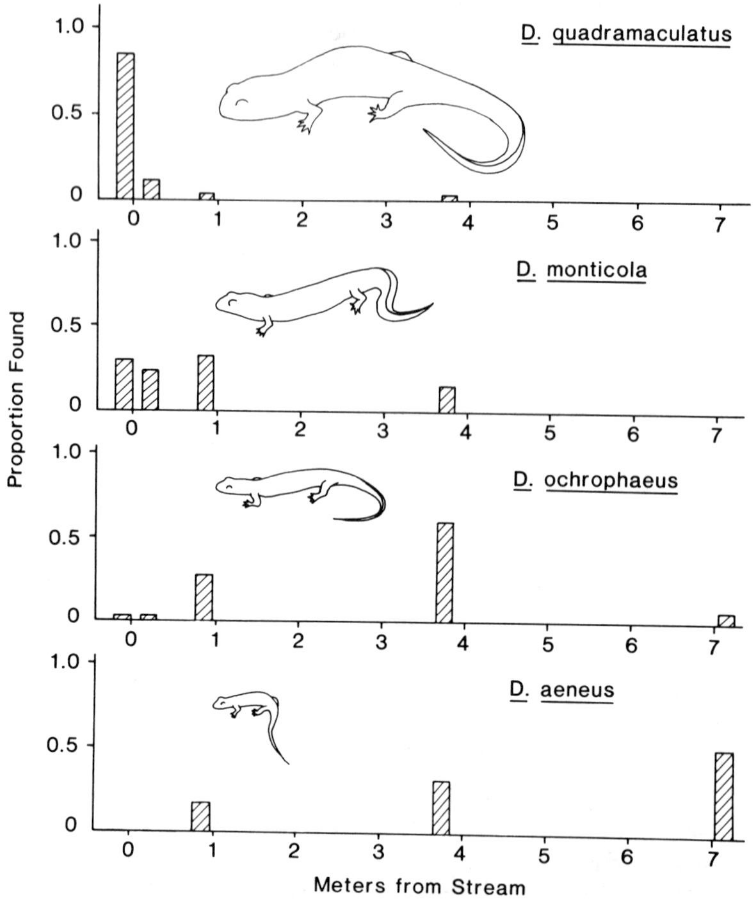

Figure 6.1. The four coexisting species of *Desmognathus* at Coweeta in the Nantahala Mountains of North Carolina. Outlines are shown to scale for adults of average size and are shown with ecological distributions.

efficiencies that should come from niche partitioning (Fig. 6.1). Lungless salamanders depend to a considerable extent on cutaneous respiration, which means that the skin must be kept moist, even at the risk of desiccation. The smaller the individual, the greater the risk, as is obvious from surface–volume relationships. Therefore, the fact that size decreases with terrestrialism suggests that some selective pressure other than competition was responsible for the series of species. In reviewing other implications of the niche-partitioning hypothesis, I found eight predictions about the ecological distribution or the size relationships of the remaining species where one of the series was absent.

Five of the predictions were rejected by observations; only three of the predictions were borne out, and I proposed a set of experiments that would allow us to choose between the competition and predation hypotheses (Hairston 1980b).

Experiments on ecological interactions in *Desmognathus*

Test for predation and competition: methods

The three species most commonly found together in the southern Appalachians are *D. quadramaculatus*, *D. monticola*, and *D. ochrophaeus*. If *D. ochrophaeus*, the smallest and most terrestrial, were removed from experimental plots, the adjacent species, *D. monticola*, would experience a benefit only if the relationship was competition. If predation was the important interaction, *D. monticola* should not gain from the removal of its smaller congener, and it and the still larger *D. quadramaculatus* might lose, if *D. ochrophaeus* was an important prey item. On the other hand, removal of *D. monticola* from other experimental plots should benefit *D. ochrophaeus*, whether the interaction was competition or predation. Removal of *D. monticola* should benefit the larger *D. quadramaculatus* only if they were competing.

The experiments were carried out in the forest operated by the Coweeta Hydrologic Laboratory of the United States Forest Service. The forest consists of numerous small watersheds that are maintained for large-scale experiments and are monitored for weather factors and for the flow of the streams draining them. Three of these watersheds were selected, and the experiments conducted in a block design, each watershed serving as a block. Three plots were staked out along the stream draining the watershed (Hairston 1986). Originally, they extended for 30 m along the stream, and 7 m laterally on both sides. The plots were separated from each other by at least 30 m. The lateral distance was chosen because preliminary observations had shown that 96 to 99 percent of the individuals of the three species were found within 6 m of streams. A fourth species, *D. aeneus*, was present, and being the most terrestrial, it was not confined to the plot area – half of the specimens seen being more than 6 m from water. On each stream, one plot was designated as the "*ochrophaeus*-removal plot," one was designated as the "*monticola*-removal plot," and one was designated as the control. Four searches were made of each plot in May and June 1981, before manipulations began, in order to provide baseline data for all treatments. All individuals of each species seen were classified as to size, but were not disturbed other than by the presence of the investigators. Searches were made at night, when salamanders are active above ground, thus avoiding repeatedly disturbing the habitat by turning logs and stones.

Experimental removals began in July 1981 and continued twice monthly through July 1982. By that time it was clear that the removals were not being effective: Removal plots had as many or more individuals of the species being removed as did the control plots. It was necessary to change the protocol so as to increase the effort. That was accomplished by reducing the size of each plot to 10 m along the stream and searching three times in succession at each visit. The new method was successful, and the experiment was continued through September 1984.

Results of experimental removals: population numbers

Effect of D. monticola *on* D. ochrophaeus. The design of the study required that a positive effect be demonstrated on *D. ochrophaeus* where *D. monticola* was removed, because that was expected under either of the two hypotheses being tested. Without that result, the whole experiment would have been a failure, as it would have been had I not been able to show that my efforts reduced the populations of the species being removed. Removing *D. monticola* did have the expected effect (Fig. 6.2). During the period of baseline observations, there was no significant difference between the numbers of *D. ochrophaeus* seen on the control plots and on the plots designated to have *D. monticola* removed. After the start of the removals, the number of *D. ochrophaeus* on the *monticola*-removal plots rose above the number on the controls and remained so on twenty-five of the remaining twenty-seven visits to the plots. During the last two years, the means were different by a statistically significant amount. In the last year, after the situation had apparently stabilized, the increase was to a mean level 41 percent above the mean on the control plots. There was also a complex effect on the ecological distribution of *D. ochrophaeus*, discussion of which will be deferred until after a description of the results of the experiments on the abundance of *D. monticola* and *D. quadramaculatus*.

Effect of D. ochrophaeus *on* D. monticola. The preliminary assignments of plots to treatments resulted in the anomalous situation that *D. monticola* was significantly more abundant during the baseline period on plots designated to become *ochrophaeus*-removal plots than it was on the plots designated to become controls. The difference amounted to 85 percent of the mean control density. Even though the mean density on the experimental plots remained higher than the mean on the controls for the next eleven visits, the difference was never significant, and during the last two years of the experiment the lines depicting the means crossed and recrossed (Fig. 6.3), indicating that the treatment had caused the densities to converge. The removal of *D. ochrophaeus* had no consistent effect on the ecological distribution of *D. monticola*. Thus, the prediction from the predation hypothesis was confirmed, and the prediction from the competition hypothesis can be rejected, as far as these two species are concerned.

Effect of D. ochrophaeus *on* D. quadramaculatus. The predictions of the outcome of removing *D. ochrophaeus* are similar for the effect on *D. quadramaculatus* to those on *D. monticola*. If the species are competing, *D. quadramaculatus* should gain; if it is a predator on *D. ochrophaeus*, removing the latter should provide no benefit, and *D. quadramaculatus* might decrease in numbers, if *D. ochrophaeus* is an important food item. Because the two species overlap less in distribution than either does with *D. monticola*, an effect might be expected to be less intense and less likely to be statistically significant than in the effect on *D. monticola*.

During the first two years of the experiment, there was no significant difference at any time between the mean densities of *D. quadramaculatus* on control plots and on those from which *D. ochrophaeus* was removed (Fig. 6.4). On two series of searches in 1983, numbers on the control plots were significantly greater than on the experimental ones, and for 1984 the mean number of *D. quadramaculatus* over the year was greater than the mean on plots from which *D. ochrophaeus* was removed by a highly significant amount.

The experiment, like the effect on *D. monticola* of removing *D. ochrophaeus*, shows that under natural conditions, *D. ochrophaeus* must have been exerting a beneficial effect on the larger species. Thus, the predation hypothesis was confirmed, and the result indicates that *D. ochrophaeus* is a significant item in the diet of both *D. monticola* and *D. quadramaculatus*. The results agree with the conclusions of Huheey and Brandon (1973) about the relative densities of *D. ochrophaeus* along forest streams and on rockfaces. It will be recalled that they argued that predation on forest populations was the reason for the marked difference. The combination of results also suggests strongly that at least one component of the interaction between the two larger species is competition for an important food item.

Effect of D. monticola *on* D. quadramaculatus. The relationship between these two species may eventually turn out to be very complex. The simple

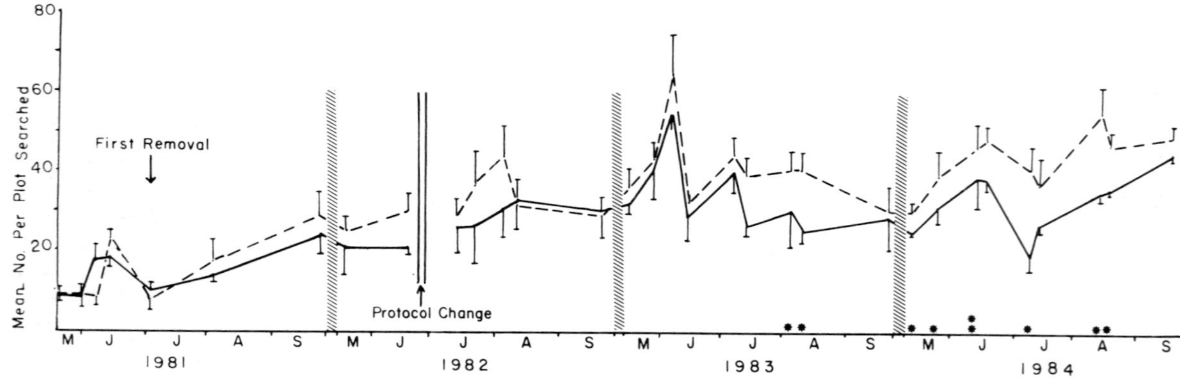

Figure 6.2. Response of the *D. ochrophaeus* population (solid line, control) to the removal of *D. monticola* (dash line). Vertical bars, 1 SE; *p < .05; **p < .01. (From Hairston 1986.)

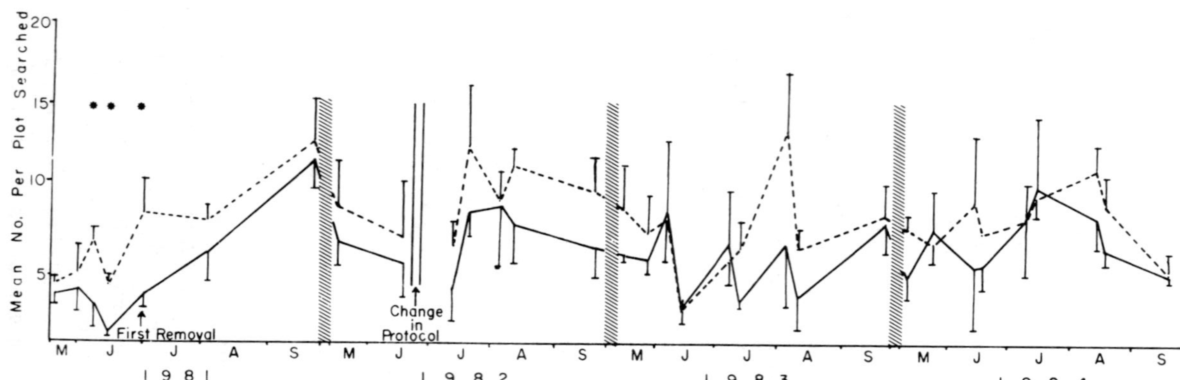

Figure 6.3. Effect on *D. monticola* abundance (solid line, control) of the removal of *D. ochrophaeus* (dash line). Vertical bars, 1 SE; *p < .05. (From Hairston 1986.)

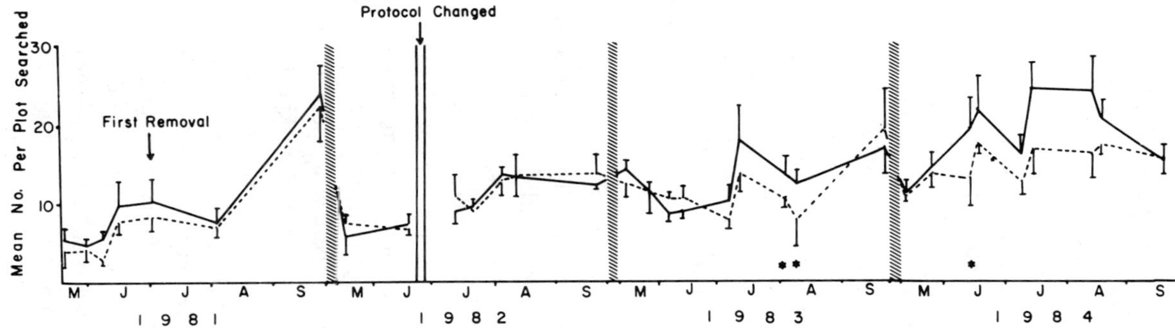

Figure 6.4. Effect on the *D. quadramaculatus* population (solid line, control) of the removal of *D. ochrophaeus* (dash line). Vertical bars, 1 SE; $^{*}p < .05$. (From Hairston 1986.)

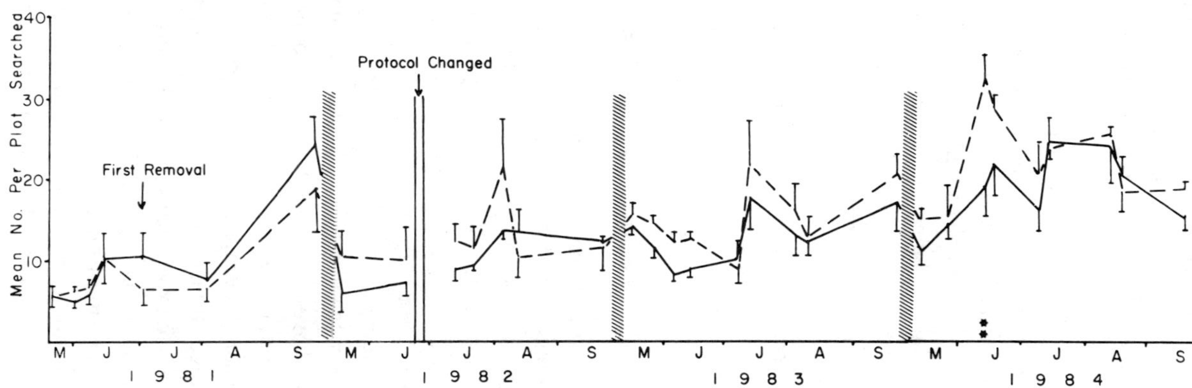

Figure 6.5. Response of the *D. quadramaculatus* population (solid line, control) to the removal of *D. monticola* (dash line). Vertical bars, 1 SE; $^{**}p < .01$. (From Hairston 1986.)

assumptions made in planning the experiment would permit a decision between predation and competition, but if both interactions are involved, the outcome of the experiment of removing *D. monticola* would depend on at least three variables: the relative abundance of the two species at the start of the experiment, the strength of competition, and the intensity of predation by *D. quadramaculatus* on *D. monticola*. With regard to the first factor, *D. quadramaculatus* was three times as abundant on control plots as was *D. monticola*. Combined with its smaller size, that suggested that any effect of removing the latter species might have been difficult to detect. Figure 6.5 shows that there was relatively little effect on the population of *D. quadramaculatus*, but that there was a significant positive effect in 1984, the last year of the experiment. Thus, if *D. quadramaculatus* is a predator on *D. monticola*, the relationship was not detectable under the prevailing conditions of relative abundance. The weak positive effect of removing *D. monticola* indicates that competition was the prevailing interaction.

Changes in ecological distribution

Effect of D. monticola *on the distribution of* D. ochrophaeus. The ecological distributions of the species of *Desmognathus*, as reported earlier (Tables 1.1 and 6.1), represent what is found when cover objects are moved during the day in summer. All species forage farther from water at night than these tables show, and there are seasonal shifts in ecological distribution as well (Organ 1961). Most individuals are found in streams, or very close to them, during the colder months. The movement away from water is gradual and is reflected in the mean distance from water at night (Table 6.2). These were specimens seen above the surface at night – no logs or stones were turned in making the observations. Adults were found farther from water than metamorphosed immature specimens, and there was an increase during the summer in the average distance from water. As noted earlier, the reactions of the other species to experimental removals were complex. It would be expected that the removal of either a predator or a competitor might result in a shift toward the vacated part of the habitat, and that phenomenon was observed in *D. ochrophaeus* for a brief period following the intensification of effort to remove *D. monticola* in 1982 (Table 6.3). Up to the start of the extra effort, there was no significant difference in the distribution of *D. ochrophaeus* on control plots as compared with its distribution on *monticola*-removal plots. The table shows that from the start of intensified effort in July, there was a marked shift toward the stream on plots from which *D. monticola* was removed. The complexity

Table 6.2. *Seasonal changes in mean distance (m) from water for three species of* Desmognathus

		Month				
Species	Age class	May	June	July	Aug.	Sept.
D. quadramaculatus	Immature	0.02	0.27	0.55	0.23	0.05
	Adult	0.54	1.03	1.38	1.63	0.20
D. monticola	Immature	0.25	1.74	1.10	2.01	0.22
	Adult	0.37	2.33	1.78	3.05	0.35
D. ochrophaeus	Immature	2.24	2.68	2.85	2.18	2.33
	Adult	2.08	3.18	2.89	3.10	2.79

Source: Hairston (1986).

Table 6.3. *Distribution of* D. ochrophaeus *before and after the start of intensified effort to remove* D. monticola*; protocol changed after June; body of table gives proportions at each distance from stream*

		Distance from stream (m)				
Month	Treatment	0–2	2–4	4–7	*N*	χ^2 (2 d.f.)
May	Control	.43	.38	.18	99	
	D. monticola removed	.41	.44	.15	105	1.17 (NS)
June	Control	.52	.22	.27	79	
	D. monticola removed	.54	.27	.19	115	4.06 (NS)
July	Control	.34	.30	.36	159	
	D. monticola removed	.45	.33	.22	204	18.79[a]
Aug.	Control	.24	.41	.34	203	
	D. monticola removed	.41	.34	.25	231	33.00[a]
Sept.	Control	.20	.39	.41	49	
	D. monticola removed	.43	.39	.17	69	27.34[a]

[a] $p < .001$.
Source: Hairston (1986).

of this situation became apparent only during the last two years of the experiment, when the ecological distribution of *D. ochrophaeus* showed no consistent pattern when control and *monticola*-removal plots were compared. This peculiarity can be understood only when the relationship between *D. monticola* and *D. quadramaculatus* is considered.

Effect of D. monticola *on the distribution of* D. quadramaculatus. There was a shift in the distribution of *D. quadramaculatus* following the increase in effort to remove *D. monticola*, and, like the change in abundance, the shift was in the expected direction, but was not found in all months. The proportion of specimens observed away from the streams was greater for *monticola*-removal plots in nine of the thirteen separate months of observation, starting in July 1982. The difference was statistically significant in five of these months. For no month was there a significant shift in the opposite direction. Thus, the competitive effect between these two species is again weakly confirmed.

Effect of D. quadramaculatus *on the distribution of* D. ochrophaeus. Most of the shift in the distribution of *D. quadramaculatus* from the water to the streambanks took place early in 1983, when *D. ochrophaeus* was found to have reversed its shift from the forest toward the streams in response to the removal of *D. monticola*. The obvious interpretation is that when *D. monticola* was replaced on the streambank by its larger congener, *D. ochrophaeus* was again at a disadvantage there, because *D. quadramaculatus* is at least as effective a predator as *D. monticola*. These are admittedly post facto explanations, not included in the original experimental design, but they are plausible and consistent with the other experimental results.

Position of D. aeneus *in the community*

Desmognathus aeneus, the smallest and most terrestrial member of this community, was not affected by the removal of either *D. ochrophaeus* or *D. monticola*. From the start of the increased effort in 1982, the total numbers seen on all plot searches combined were twenty-one on the *ochrophaeus*-removal plots, fourteen on the *monticola*-removal plots, and twenty on the controls. These are much smaller than the numbers of the other species seen on control plots alone, and they probably reflect a combination of effects. First, as described earlier, *D. aeneus* had a distribution that extended well beyond the limits of the plots – more than half of the population had been found more than 6 m from streams. Second, unlike any other salamander species with which I am familiar, it comes to the surface at night in very low numbers. During the early part of the experiment, plots were searched both during the day and at night. During the daytime searches, easily moved cover objects were turned and carefully replaced. I saw 3.75 times as many *D. quadramaculatus* at night as during the day, 1.81 times as many *D. monticola*, and 1.65 times as many *D. ochrophaeus*. In contrast, I found six times as many *D. aeneus* during the daytime searches as at night. Thus, the technique adopted (night searches) would have minimized the estimate of the density of *D. aeneus*. It is clearly a secretive member of the community, and unlike the others, especially *D. ochrophaeus*, it was never seen climbing on the trunks of shrubs and trees.

Desmognathus aeneus is so different from its coexisting congeners in ecological distribution and in behavior that the experimental removal of neither *D. monticola* nor *D. ochrophaeus* had a measurable effect on its population. There is geographic variation in the behavior of *D. aeneus* that is described as part of the general discussion of the evolution of desmognathine salamanders in Chapter 7.

Complexity of the ecological interactions in the Coweeta Desmognathus *community*

These experiments have shown that predation is the most clearly demonstrable interaction among the species. Of the six pairs of species, two were shown to involve predator–prey relationships, one pair involved competition, and two pairs involved no detectable interaction. For six pairs of species and three possible kinds of relationships there are eighteen combinations, of which only four were not tested, or not tested adequately: predation for *D. quadramaculatus* and *D. monticola*, and the three relationships that could have involved *D. quadramaculatus* and *D. aeneus*. Of the remainder, the seven not shown in Figure 6.6 were eliminated by the experimental results. In the case of *D. quadramaculatus* and *D. monticola*, the latter was shown to have a negative effect on the former. Inasmuch as *D. quadramaculatus* is the larger of the two, the effect is almost certainly competition. No experiments have been designed to test for the possibility that predation is involved. The removal of *D. quadramaculatus* would probably give a positive result. That result, however, would permit a distinction to be made between competition and predation only if *D. quadramaculatus* did not respond to the removal of *D. monticola*. I have observed this largest member of the community in the act of eating another large desmognath, but the latter was too mangled for identification, and I may have witnessed cannibalism, a relationship that I have observed in both *D. monticola* and *D. ochrophaeus* in captivity. Southerland (1985) has watched natural examples of predation by *D. quadramaculatus* on both *D. monticola* and *D. ochrophaeus*, as well as a field example of cannibalism by *D. monticola*.

Superficially, *Desmognathus* provides an excellent example of a guild, as defined by Root (1967). The species subsist on overlapping arrays of prey

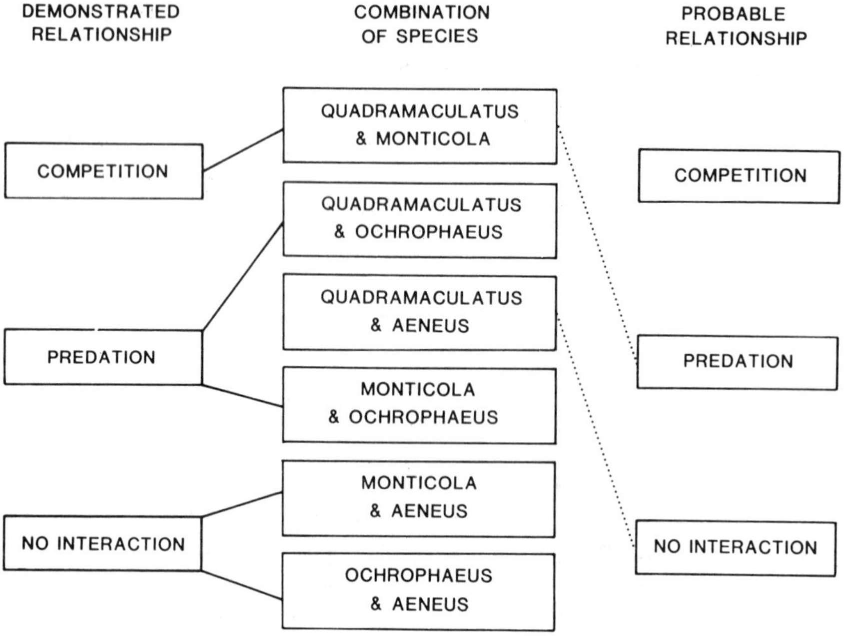

Figure 6.6. Ecological interactions among the four species of *Desmognathus* at Coweeta Hydrologic Laboratory. Known relationships are shown with soild lines; probable relationships with dotted lines; unlikely but possible interactions (e.g., competition between *D. quadramaculatus* and *D. aeneus*) are omitted.

species (Hairston 1949; Krzysik 1979); they occupy different but overlapping ecological distributions; and they are of sufficiently different sizes to suggest that they have reduced niche overlaps enough to permit coexistence. The failure to detect any effect of *D. monticola* or *D. ochrophaeus* on *D. aeneus* might be explained as an example of "the ghost of competition past" (M. Rosenzweig: see Connell 1985), but the combination of predation and competition within the guild is not predictable from any community theory with which I am familiar. This example points once again to the risk of accepting theory based on observational evidence without insistence on a priori predictions. These most frequently require experimentation – a method of making certain that the a priori requirement is met.

Mechanisms of interspecific interaction among desmognathines

The behavioral, ecological, and physiological mechanisms by which these salamanders react to the interspecific relationships that confront them have been investigated from a number of standpoints. The experiments have been conducted in Pennsylvania, Ohio, Virginia, and North Carolina. The combinations of species naturally present were *D. fuscus* and *D. ochrophaeus*, *D. fuscus* and *D. monticola*, *D. monticola* and *D. quadramaculatus*, and the four-species combination of *D. quadramaculatus*, *D. monticola*, *D. fuscus*, and *D. ochrophaeus*. The results have frequently, but not always, depended on relative size, as when *D. monticola* was paired with *D. fuscus* on one hand, or with *D. quadramaculatus* on the other.

Behavioral responses to congeners or conspecifics

The most immediate response available in reacting to the presence of another individual is in altering behavior, and overt aggression is easily interpretable. The behavior that is recorded may be secondary to aggression, which may not have immediately preceded what was recorded, but could have been part of the responding individual's previous experience. Male-to-male aggression within species is common among captive specimens during courtship season. Organ (1961) reported attacks with biting among males of *D. quadramaculatus*, *D. monticola*, *D. ochrophaeus*, and *D. wrighti*. Direct observation of aggression in the field has been reported once (Brandon & Huheey 1971): A specimen of *D. monticola* escaping from the observers entered the hole occupied by a larger conspecific individual. It was attacked and driven out of the hole. Interspecific aggression was observed by Keen and Sharp (1984) in small laboratory arenas that had been occupied by a "resident" specimen of *D. monticola* for several days. Intruding individuals of either *D. monticola* or *D. fuscus* were attacked in 101 of 204 trials. As with Organ's observations, relative size was important, with the smaller specimens being attacked. Keen and Sharp suggested that with a sufficient size difference, aggression and predation become indistinguishable – an important consideration in a genus as divergent in size as *Desmognathus*. Southerland's field observations (1985) of predation and cannibalism, mentioned earlier, are relevant in this respect.

Despite the apparent importance of aggression in plethodontid ecology, direct observations are rarely recorded, and behavioral responses to other individuals most frequently are of kinds that could be an indirect result of aggression. Such responses have frequently involved changes in ecological distribution. The observations have been of correlative kinds ("natural experiments") and of the results of manipulative experiments. A striking example of the difference in ecological distribution of a species in the presence of congeners and in their absence was provided by Southerland (1986b). *D. fuscus* is the sole representative of the genus on the piedmont of North Carolina. In the

habitat selection in the presence of congeners ↓ historical effects

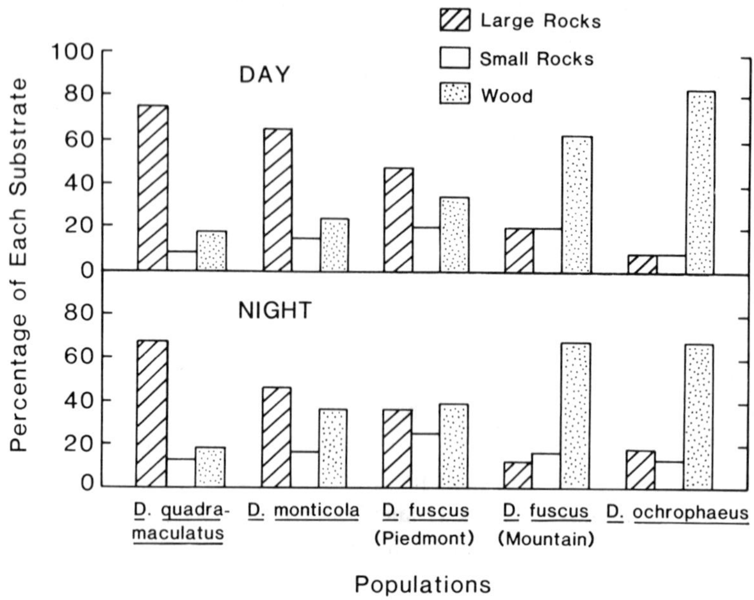

Figure 6.7. Patterns of substrate choice for five populations of *Desmognathus* salamanders. Individuals from four sympatric populations of congeneric species were collected from the same mountain site (Boone, North Carolina), whereas individuals from a fifth population were collected from a piedmont site (Lenoir, North Carolina) where congeners do not exist. Each column presents the percentage of each substrate chosen by thirty-six individuals from each population. Day and night observations are shown separately. (From Southerland 1986b.)

Blue Ridge Mountains, there are four more species, three of which coexist with *D. fuscus* in some areas. Southerland conducted experiments testing for the choice of each of the four coexisting species among three kinds of substrates imitating the array available in the gradient from aquatic to terrestrial habitats: large (50 cm^2) rocks, small (10 cm^2) rocks, and small (10 cm^2) pieces of wood. The percentage of individuals on each kind of substrate matched the observed relative distributions along the aquatic–terrestrial gradient (Fig. 6.7), with *D. quadramaculatus* showing the greatest affinity for large rocks, and *D. ochrophaeus* the greatest affinity for wood. *D. monticola* was similar to *D. quadramaculatus*, but less attracted to large rocks, and local *D. fuscus* was intermediate between *D. monticola* and *D. ochrophaeus*, but more like the latter in substrate choice. *D. fuscus* collected at a piedmont locality only 50 km away was almost indiscriminate in its choice of substrate. Each salamander was tested separately, and the conditions were such that no direct influence by other specimens was possible. Although there is only indirect evidence of genetic differences between these populations of *D. fuscus* (Karlin, Guttman, & Rathbun 1984), the implication is strong that the species has responded in an evolutionary way to the presence (or absence) of congeneric species. The mountain form has "moved" away from the larger species, which are most likely to be predators, and toward the smaller species, which is most likely to be a competitor. For spatial distributions in an artificial stream in the laboratory, Carr and Taylor (1985) observed mutual negative effects between *D. monticola* and *D. ochrophaeus*. *D. ochrophaeus* had a negative effect only on juvenile specimens of *D. monticola*. Shealey (1975), however, found no differences in the patterns of seasonal activity of these two species. At the elevation of his observations (473 m), both were maximally active during April. The only deviation was in mature males of *D. ochrophaeus*, which showed a peak in July. Both temperature and precipitation had immediate effects on nocturnal activity. The observations are mostly consistent with those of Stewart and Bellis (1970), who found both *D. fuscus* and *D. ochrophaeus* to be randomly distributed under carefully placed boards along a stream in Pennsylvania. They also agree with with those of Holomuzki (1980), who found no difference in the pattern of nocturnal activity between the two species in northeastern Ohio.

Experiments on pairs of species of *Desmognathus* have shown significant effects on the behavior of the smaller species. Keen (1982) observed that *D. fuscus* in field enclosures 290 × 80 cm chose microhabitats significantly farther from a stream when in the presence of *D. monticola* than when alone, and Southerland (1986a) found that juvenile *D. monticola* shifted their choice of substrate away from wood to large rocks in the presence of adult *D. fuscus*, with the distribution of the latter becoming more uniform than it was when alone. Keen, however, did not observe an effect of either species on the choice of any aspect of the substrate by the other, when they were confined in laboratory arenas. He did observe that the nocturnal activity of *D. fuscus* was reduced in the presence of *D. monticola*.

In the presence of the larger *D. quadramaculatus*, *D. monticola*'s role is sometimes reversed. In Southerland's experiments (1986a), both adults and juveniles shifted their choices of substrate away from large rocks, although Keen (1985) did not observe such differences. Southerland also observed that the nocturnal activity of *D. monticola* in large field enclosures was reduced to 50 percent of the control amount by the presence of *D. quadramaculatus*. Carr and Taylor (1985) observed the same effect in their artificial stream.

The combination of aggression and the observed shifts in habitat and substrate suggests that there should be some form of avoidance, despite the negative finding of Stewart and Bellis (1970). Southerland (1986a) conducted an extensive set of experiments using multiple cover objects and various pairs of adult and juvenile specimens of *D. monticola*, *D. fuscus*, and *D. ochrophaeus*. Individuals in all combinations were observed together under the same object less frequently than would be expected by chance, and the total cooccurrence (15 percent) was significantly below the expected 35 percent. Intraspecific and interspecific avoidances did not differ significantly. He also searched the bank of a stream where both *D. quadramaculatus* and *D. monticola* were abundant and measured the distances between the burrows of nearest neighbors (Southerland 1985). There was a significant deficiency of short distances, with no pairs less than 20 cm apart. Again, there was no difference between intraspecific and interspecific pairs. He agreed with Brandon and Huheey (1971) that streambank refuges, especially burrows, form an important resource as a safe location from which to forage, and he noted that all of the burrow-holders were adults and suggested that juveniles of both species must resort to using stones or other less desirable cover objects along the streambank.

Physiological responses to the presence and density of conspecifics and congeners

Like other traits, the physiology of an organism may respond to stress or to opportunity, and the response may be direct or may reflect the evolutionary history of the species. In the latter category are the relative rates of absorbing yolk and of development by the embryos for two pairs of species of *Desmognathus* (Orr & Maple 1978). They were *D. fuscus* and *D. ochrophaeus* in northern Ohio, and *D. quadramaculatus* and *D. monticola* in North Carolina. In both pairs, the member with the larger adult size absorbs yolk faster and hatches earlier that its smaller congener. Orr and Maple interpreted their observations to mean that the interspecific differences give increased size differences between the newly hatched larvae, with the result that competition for similar-sized prey is reduced. They compared the development of *D. fuscus* from locations sympatric and allopatric with *D. ochrophaeus* and found that they differed from each other in the predicted direction.

Storage of fat is widely regarded as an index of successful existence (Maiorana 1977). Thus, as an effect of competition, an individual might suffer a loss of fat. Kleeberger (1984) investigated this possibility by increasing the densities of conspecific and congeneric specimens of *D. monticola* and *D. quadramaculatus* in unfenced plots in an area near Highlands, North Carolina. In a separate experiment, he added rocks of appropriate size, as well as additional specimens of *D. monticola*. The addition of *D. quadramaculatus* was unsuccessful in raising its density, and there was no effect of this treatment on either the density or the total lipid content of *D. monticola*. With the latter species, the manipulations were successful. Additions increased the number found on subsequent searches of the plots, and there was a significant decrease in the total lipid content of such individuals at the end of the experiment below that of animals from the control plots. Thus, direct crowding reduced their ability to store fat. The addition of rocks for refuges along with extra individuals of *D. monticola* had the opposite effect: an increase in the total body lipids above that found in control animals. Whether extra effort in searching for refuges or a reduction in available food per salamander was responsible for the reduced lipids under crowded conditions cannot be determined, but the reversal of results following addition of rocks suggests that food was not responsible, because salamanders were also added. The further evidence that crowded *D. monticola* had significantly larger home ranges than controls (Kleeberger 1985) strengthens this view. Kleeberger concluded that there was no evidence from his study for interspecific competition.

A gain or loss of weight can be a sensitive indicator of the effect of competition. Southerland (1985) followed the weights of individual *D. monticola* in four pens 5 m^2 in area under the following densities in each pen: five and ten conspecifics; five conspecifics plus five *D. quadramaculatus* plus five *D. ochrophaeus*; ten conspecifics plus ten *D. ochrophaeus*; and ten conspecifics plus ten *D. quadramaculatus*. In the first two of these treatments, there were increases in weight by most individuals and a gain in average weight that was comparable to that in free-ranging specimens (Table 6.4). The *D. monticola* confined with five of each of its congeners showed less increase in average weight over the ten weeks of the experiment, and there was a loss of weight at the two highest densities, which included *D. ochrophaeus* or *D. quadramaculatus*. These are not completely unrealistic densities for the three species, but they obviously exceeded the carrying capacity under the experimental conditions. Southerland concluded that the results were consistent either with reduction in prey or with interspecific interference during foraging.

Populational responses to conspecifics and congeners

Changes in population density may be brought about through the movement of individuals into or out of a specificed area, but it is more conventional to associate such changes with changes in reproduction or survival. I recorded increases in reproduction by *Plethodon jordani* following the

Table 6.4. *Mean growth measures of* Desmognathus monticola *under different densities of conspecifics and congeners*

Treatment	Density (m^{-2})	*N*	Mass increase[a] (%/week)	Specific growth rate[b] Mean	SD
1982					
5 DM[c]	1	20	2.6	0.011	0.010
10 DM	2	40	3.7	0.014	0.029
10 DM, 10 DO	4	20	−7.0	−0.018	0.025
10 DM, 10 DQ	4	20	−9.2	−0.023	0.028
1983					
5 DM	1	20	5.0	0.015	0.010
5 DM, 5 DQ, 5 DO	3	20	2.7	0.010	0.006
Free-living DM	—	24	2.2	0.010	0.018

[a]Relative to initial body mass in grams.
[b]Calculated as [log(final mass) − log(initial mass)]/no. weeks.
[c]DM = *D. monticola*; DQ = *D. quadramaculatus*; DO = *D. ochrophaeus*.
Source: Southerland (1986c).

removal of its competitor *P. glutinosus* (see Chapter 5), but similar observations have not been made on any species of *Desmognathus*, the reports being confined to survival. Hall (1977) followed marked individuals of *D. fuscus* and *D. ochrophaeus* in northern Pennsylvania and found that the former had an appreciably lower rate of survival. *D. fuscus* also had a higher proportion of broken-off tails, and Hall concluded that this observation suggested a much greater predation pressure in the more aquatic habitat of *D. fuscus*. In the area where he worked, the larger *D. monticola* is absent, and predation would have to be by some other species. Hall's observations are reminiscent of Organ's analysis (1961) of comparative survival for the series of five species of *Desmognathus*, in which there was a regular progression of improving survival with position on the aquatic–terrestrial gradient (Fig. 3.8).

Southerland (1985) conducted experiments to compare the survival of *D. monticola* under different conditions of habitat type, intraspecific density, potential interspecific competition, and potential predation. Cages 0.25 m^2 in area were placed in forest-floor, streambank, and stream locations, and two specimens of *D. monticola* were placed in each cage. Comparison between cages in different locations revealed the effect of habitat. Matched cages in each location also had two specimens of the appropriate species for the habitat: forest, *D. ochrophaeus*; streambank, *D. fuscus*; and stream, *D. quadramaculatus*. In each of these cages, the potential competitors were of the same size as the *D. monticola*, as were two additional specimens in a streambank cage for the intraspecific crowding effect. Additional cages on the streambank and in the stream had large *D. quadramaculatus* added as potential predators. Twelve replicates were carried out sequentially, with rotation of treatments among cages at the same location. There was a significant effect for all treatments on the survival of *D. monticola*. In the streambank experiments, there was no difference between the effect of intraspecific crowding and crowding by *D. fuscus*. Survival was best (but poor) away from the stream, and worst in the presence of predatory *D. quadramaculatus*. All competing species survived throughout the experiments (up to 109 days), but *D. monticola* survived less than thirty days in all conditions, and ten days or less except under the best conditions. This was the same experiment described earlier in which changes in weight were followed, and survival was inversely correlated with weight loss.

There have been two experiments testing the effect of adding cover objects on the density of *Desmognathus* species. Kleeberger's experiment (1984) in adding rocks resulted in a significant increase during the summer season in the number of *D. monticola* to 2.6 times the initial number, compared with an increase of 8 percent in the control plots.

Southerland (1985) established twelve 1.0-m^2 streambank plots at each of two locations in North Carolina, at only one of which was *D. fuscus* present. After four searches of each plot to establish baseline data, he added rocks to one-third of the plots and pieces of wood to another one-third. The added cover raised the amount to approximately three times the initial density. The salamander fauna consisted almost exclusively of species of *Desmognathus* (715 of 731 specimens captured). The plots were searched regularly over a period of sixteen weeks, and both kinds of the added cover resulted in significant increases in salamander numbers (Fig. 6.8). The abundance of *D. monticola* increased significantly with both kinds of added cover, but *D. quadramaculatus* responded only to added rocks, whereas *D. ochrophaeus* increased significantly only on plots to which wood had been added.

Synthesis — V. good

Salamanders associated with the borders of streams respond to the environmental gradient of moisture, a feature that provides the possibility of easily studied ecological differences. Several genera are found in such situations, but interspecific interactions have been studied only among the mem-

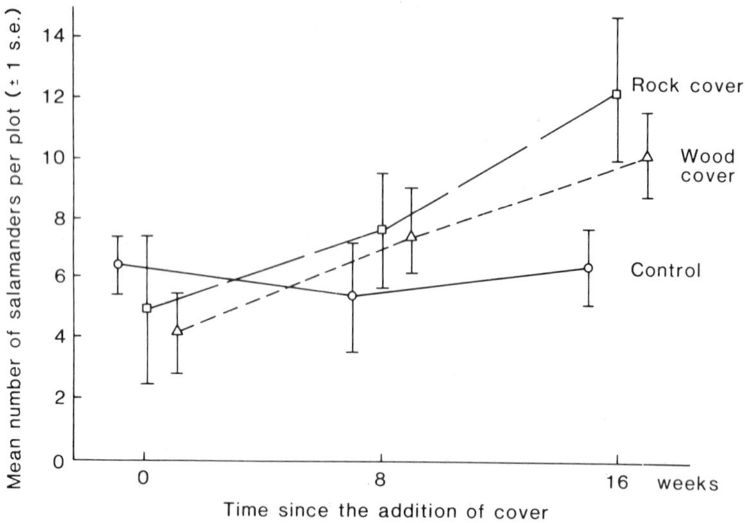

Figure 6.8. Abundances of all salamanders captured on streambank plots at Coweeta Hydrologic Laboratory, North Carolina. Salamander numbers (summed over two day and two night searches on four plots) represent the baseline sample (0 weeks) and two subsequent samples after cover was increased threefold on cover (rocky and woody) treatment plots. (From Southerland 1986d.)

bers of the Subfamily Desmognathinae of the eastern United States. Attention was first directed to interspecific differences in the type of stream inhabited, but the most reliable data came from recording the position of each animal with respect to the distance it was found from water. There is a relationship between the diet of a species and its position on the aquatic–terrestrial gradient, but a disproportionate number of food items had come from terrestrial habitats. The gradient is also accompanied by a regular decrease in adult body size as the series progresses toward the terrestrial end. These differences were originally interpreted as reflecting evolution under the influence of interspecific competition.

Doubts began to be expressed when it was realized that the size gradient was in the wrong direction, because surface–volume relations lead to the expectation that wet-skinned animals should be larger the more terrestrial they are. This would reduce the relative loss of water. An alternative hypothesis was proposed, suggesting that the high level of predation in and near streams would put selective pressure on the smaller species in the series to become more terrestrial. A set of experiments was proposed, the purpose of which was to permit a choice between the hypotheses of competition and predation in structuring the community of the species of *Desmognathus*, and six years later the results were published. Removal of the smallest of three species caused a reduction in the abundance of both of the two larger species, relative to controls. This result strongly supports the predation hypothesis. Where the species of intermediate size was removed, the smallest of the three responded by becoming significantly more abundant than controls, a result that is required by both hypotheses. The largest species also responded favorably to the removal of the intermediate species, however – a result that is best interpreted as showing the existence of competition. Finally, a fourth species, the smallest and most terrestrial one present, was not affected by the experimental removals of its larger congeners. Thus, all three possible results were obtained, depending on the combination of species involved, although predation appears to be the most important interaction in the community.

The ecological relationships that have been demonstrated among the species of *Desmognathus* can best be understood from a number of experiments involving the manipulation of individuals and of the type and amount of cover that was made available to them. Aggression is the basis of practically all of the findings in these experiments, and aggressive defense of burrows and other refuges accounts for the results. In the presence of a larger congener, each of several species shifted its position away from the preferred position in the gradient of habitats, shifted away from its preferred kind of cover object, and reduced its amount of surface activity. One species, with one population cooccurring with three others and another population 50 km away living by itself, was found to have a much more restricted choice of habitat with congeners than alone. It avoided the habitats most used by the large, predaceous species. Direct aggression is rarely observed in the field, but is readily seen under confined conditions, with the larger individual always being the aggressor. Predation by larger species on smaller ones has been watched in the field on several occasions.

In experiments in which two or three species were confined in field enclosures at realistic densities, the species of intermediate size suffered increased mortality and decreased growth. The effect of increased cover was shown in another experiment where individuals stored more lipids when additional rocks were provided, even though extra animals were added to the system. Without the extra rocks for cover, the added animals caused a decrease in lipid storage.

In two experiments, increased densities of *Desmognathus* have been obtained by adding either rocks or pieces of wood to natural plots along streams.

In the great majority of studies, aggressive interference both with conspecific individuals and with congeners was the mechanism by which the results were brought about. It is clear that among these species, the difference between aggressive interference and predation depends on the relative sizes of the individuals involved. Because the array of species is one of decreasing size with increasing terrestrialism, the role of predation in structuring the community is evident.

7

Evolution in salamander communities: experimental tests

Introduction

Salamanders provide the exciting possibility that they can be used experimentally to reveal in a truly a priori way the evolutionary responses of species to the conditions of competition or predation to which they have been subjected in the past. It has been possible to test some of the conclusions from the large amount of theoretical work that has been done under the general topic of "evolutionary ecology." Conclusions about the evolutionary process are nearly always reached by a series of observational and logical steps known loosely as "weak inference." That is, plausible explanations are offered for information that has already been obtained. This work has concentrated on interspecific competition as the ecological basis of natural selection. The many observations of differential utilization of common resources by otherwise similar species have led to the general acceptance of mathematical theories purporting to show that natural selection will favor niche partitioning by competing species. Two works frequently cited as making that conclusion are Lawlor and Maynard Smith (1976) and Roughgarden (1976). It is not always easy to identify all of the assumptions that are made in such mathematical work, as many are contained within the mathematics, but the papers cited earlier can be dissected to yield a number of them.

Lawlor and Maynard Smith (1976) made six important assumptions:

1. The abundance of each of two competing species is completely determined by the rate of supply of each of two renewable resource species, the rates at which each consumer captures each resource, and the value of each resource to each consumer.
2. The respective rates of capture can evolve.
3. The rate of change in abundance of each kind of resource is determined in a strict density-dependent manner, involving only the density of the resource, the number of consumers, and the rates of consumption.
4. The rates at which each consumer species captures the two resource

species are negatively related: An increase in the rate of capture of one resource entails a reduction in the ability to capture the other.
5. The resource species do not compete, nor do they evolve.
6. There is an initial difference between the two consumer species in their respective rates of capture of the two resources.

There may be more assumptions that are not obvious to the mathematically unsophisticated ecologist, but these will suffice to demonstrate the problems with the approach. Assumptions 2 and 4 seem intuitively reasonable; assumptions 1, 3, and 5 do not, and assumption 6 requires comment. The ones to which I object appear to have been made in the interest of mathematical tractability, and only assumption 3 is discussed with regard to the robustness of the conclusions to violations of the assumption, as discussed later.

With regard to the assumption of an initial difference, that might easily be the case if two recently reunited species had been isolated in different parts of a steep ecological gradient or otherwise had been required to adapt to different conditions. Given the other assumptions, natural selection might well be expected to increase the difference in resource utilization, but a requirement of an initial difference of the appropriate kind imposes a severe restriction on the applicability of the conclusions.

Lawlor and Maynard Smith considered the extent to which fluctuations in the relative rates of change in abundance of the resource species affect the results of their analysis, in violation of assumption 3. These fluctuations are most simply explained as being due to stochastic events. The model predicts that so long as the magnitude of the fluctuations is slight enough to keep the relative rates of supply of the resouces between the respectively different rates at which the consumer species capture the two resource species, the prediction of resource partitioning is valid. Fluctuations beyond that limit will result in the extinction of one of the consumer species. The model is thus sensitive to such fluctuations during the very period in which it is most interesting – immediately after two competing species have come into contact, and presumably are still similar (though not identical, per assumption 6). If they are already different enough to coexist, the whole model is irrelevant to evolution under competition.

Roughgarden's treatment (1976) of the problem contains explicit statements of the assumptions made:

1. "The competition model is the Lotka–Volterra competition system" (p. 403);
2. ". . . the competition coefficient between two species is a function of their niche separation" (p. 403);
3. ". . . reciprocal competition coefficients are equal" (p. 404);

4. ". . . the carrying capacity of species-i is a function of its niche position. . . . Thus the model consists of a set of species with niche positions, x_i, and a common niche width, w. Both the competition between species and the carrying capacity of each species are inferred from the set of x_i and w" (p. 404);
5. ". . . these remarks are intended to justify our *beginning* with the evolution of niche positions while regarding niche widths as fixed" (p. 405).

Thus, Roughgarden starts his analysis with an assumption that other authors (e.g., Dawkins & Krebs 1979) have claimed is his conclusion. His real conclusions have to do with derivative issues that are not of concern here.

Neither Lawlor and Maynard Smith (1976) nor Roughgarden (1976) attempted to produce a model that would allow a consideration of the alternative possibilities of the evolution of resource partitioning and the evolution of interspecific interference mechanisms, as proposed by Pimentel and associates (1965) and Pimentel (1968). The former authors stated that they would do so, but they only waived the issue aside as being outside the problem that they were considering. Roughgarden did not cite Pimentel's paper. The issue deserves much more attention than it has received. Interference competition is mentioned routinely in discussions of interspecific interactions, and its ecological effect has been modeled by Case and Gilpin (1974) and Schoener (1976, 1978), but its evolution is a neglected field. A partial exception is Levin (1971), who modeled the evolution of competition coefficients in asexual species. The applicability of his work to interference versus consumptive competition is limited.

My concern with this problem has been in visualizing the intraspecific results of genotypic differences in resource utilization when the species is in a situation where it is limited in abundance by an array of renewable resources. I address the issue raised by Lawlor and Maynard Smith. If these resources vary in their rates of supply, as is the case among insect species being preyed on by vertebrate predators, it is unlikely that any one prey species, or even a limited combination of them, will be readily available continuously. Thus, a genotype tending to specialize among them will frequently be at a disadvantage vis-à-vis a generalist genotype. Such an effect would counteract any tendency toward resource partitioning, should the predator species be in competition with other species for the array of resources. Intraspecifically, such a situation seems likely to lead to territoriality, and interspecifically, it might well be profitable to evolve interference mechanisms. Schoener (1974a) has pointed out how the situation might lead to a partitioning of habitats, rather than of consumable resources.

Interference is a common phenomenon in competing species. Well-known examples are antibiotics, an obvious interference mechanism among competing decomposers, and interspecific aggression, documented for terrestrial salamanders in Chapter 5 and later in this chapter. Interference stands in contrast to simple relative ability to obtain a resource. Schoener (1983) has analyzed a large number of experiments in which he considered that competition had been demonstrated in field experiments. Of 171 cases, only 71 involved competition for consumable resources; 79 could be classified as involving interference, and 21 cases of "preemption" could be classified either as competition for a nonconsumable resource or as interference. From Schoener's analysis, the importance of interference is clear. Resource partitioning should not be assumed in cases of competition, and I describe later two examples in which experimentation led to its rejection as the mechanism by which observed ecological differences were produced.

Evolution under competition in terrestrial salamanders

Large plethodons

The original description of the variation in altitudinal distribution of *Plethodon jordani* and *P. glutinosus* on different mountain ranges was accompanied by an explanation that competition is more intense where the altitudinal overlap between the two species is least (Hairston 1951). This narrow overlap occurs in the Black Mountains of North Carolina and in the Great Smoky Mountains of North Carolina and Tennessee, in contrast to a wide altitudinal overlap in the Balsam Mountains of North Carolina. Only later was it appreciated that the difference, combined with geographic variation in morphology of *P. jordani*, provided an opportunity to test experimentally the hypotheses of niche partitioning versus the evolution of competitive ability in the form of interference (Hairston 1973). The experiments testing the hypothesis that competition was of different intensity in different mountain ranges are described in Chapter 5, as are the attempts to demonstrate a difference in partitioning of the resources of food or foraging microhabitat. The failure to find a greater difference in the use of these resources where competition is less intense was taken as evidence that evolution had not led to the partitioning of a consumable resource, or of microhabitats.

In the meantime, it had already been suggested that more intense interspecific competition was the result of an evolutionary increase in competitive ability and that this had resulted in a more sharply defined altitudinal separation in the Great Smoky Mountains, where that evolutionary development had occurred. An experimental test of that hypothesis was proposed as follows: *P. jordani* is recognizably different on the two mountain ranges, having red cheeks in the Smokies and gray cheeks in the Balsams. Reciprocal transfer experiments were therefore feasible, because the survival and reproduction of transferred individuals could be monitored in the new environment, thereby controlling for the technical success of the experiment. Removal of the local population of *P. jordani* and replacement with specimens from the other mountain range would expose both the transferred *P. jordani* and the local *P. glutinosus* to new competitive environments. Transfer of *P. glutinosus* was not feasible because there was no morphological marker by which they could be followed. On other plots in each mountain range, both species were removed, and foreign *P. jordani* were introduced, so as to control for the noncompetitive effect of the new environment. These plots incidentally permitted the quantitative assessment of the effect of the local *P. glutinosus* on the introduced *P. jordani*.

The whole design was thus intended to show the effect of each species in both mountain ranges on the other species from the reciprocal mountain range. The effect of *P. jordani* from the Balsam Mountains (less competition) on *P. glutinosus* from the Great Smoky Mountains would be shown in the Smokies plots where Balsams *P. jordani* were substituted for the local form. The effect of *P. glutinosus* from the Balsams on *P. jordani* from the Smokies would be shown by comparing the performance of the latter on two kinds of plots in the Balsams: those from which both local species had been removed and those on which *P. glutinosus* remained. The reciprocal experiments would show the effects of both species from the Smokies (intense competition) on the opposite species from the Balsams.

Selection for improved competitive ability, named alpha selection by Gill (1972, 1974), would be shown by a strong negative effect of *P. jordani* introduced from the Smokies on *P. glutinosus* in the Balsams. In the absence of such an effect, selection for divergence of ecological requirements would be shown by improved performance of *P. glutinosus* in the Smokies on plots where Balsams *P. jordani* had been substituted for the local form. Table 7.1 gives the design and the interpretations of the possible outcomes, as proposed before the experiments were begun (Hairston 1973).

The experiments were continued for eight years on plots in the Balsam Mountains, because it proved difficult to obtain a sufficiently dense population of red-cheeked *P. jordani* there. When success was achieved, there was a definite negative effect on the local population of *P. glutinosus* (Fig. 7.1), showing that the population introduced from the area of intense competition had a stronger effect than did the local form of *P. jordani*. Introduced speci-

Table 7.1. *The plan and possible outcomes, with their interpretations, of experiments involving reciprocal transfers of* Plethodon jordani *between areas of wide altitudinal overlap (weak competition) and narrow altitudinal overlap (strong competition); omitted are general control (no manipulation) and some very unlikely outcomes*

Manipulation	Outcome	Interpretation
In Balsam Mountains (wide altitudinal overlap)		
A. Remove *jordani*, replace with *jordani* from Smokies	1. Persistence of moved *jordani*; decrease in abundance of *glutinosus*	1. Means that introduced *jordani* had evolved a specific competitive mechanism against *glutinosus*
	2. Moved *jordani* persists; no change in *glutinosus*	2. Means that local *glutinosus* has evolved ecological differences to coexist with all *jordani*
B. Remove *jordani* and *glutinosus*, replace with *jordani* from Smokies	1. Persistence of moved *jordani*	1. Means that moved *jordani* can survive in remote area
	2. Increase of moved *jordani* greater than A.1 or A.2	2. Means that local *glutinosus* has evolved competitive ability against all *jordani*
In Great Smoky Mountains (narrow altitudinal overlap)		
C. Remove *jordani* replace with *jordani* from Balsams	1. Persistence of moved *jordani*; increase or stability of *glutinosus*	1. Means that introduced *jordani* has evolved ecological differences to coexist with all *glutinosus*
	2. Persistence of moved *jordani*; decrease of *glutinosus*	2. Means that Balsams *glutinosus* has evolved to coexist with all *jordani*, especially strong with A.2
D. Remove *jordani* and *glutinosus*, replace with *jordani* from Balsams	1. Persistence of moved *jordani*	1. Means that moved *jordani* can survive in remote area
	2. Increase of moved *jordani* greater than C.1 or C.2	2. Means that local *glutinosus* has evolved a specific competitive mechanism against all *jordani*

Sources: Adapted from *Breviora,* No. 414, and *Evolution,* Vol. 34, No. 3.

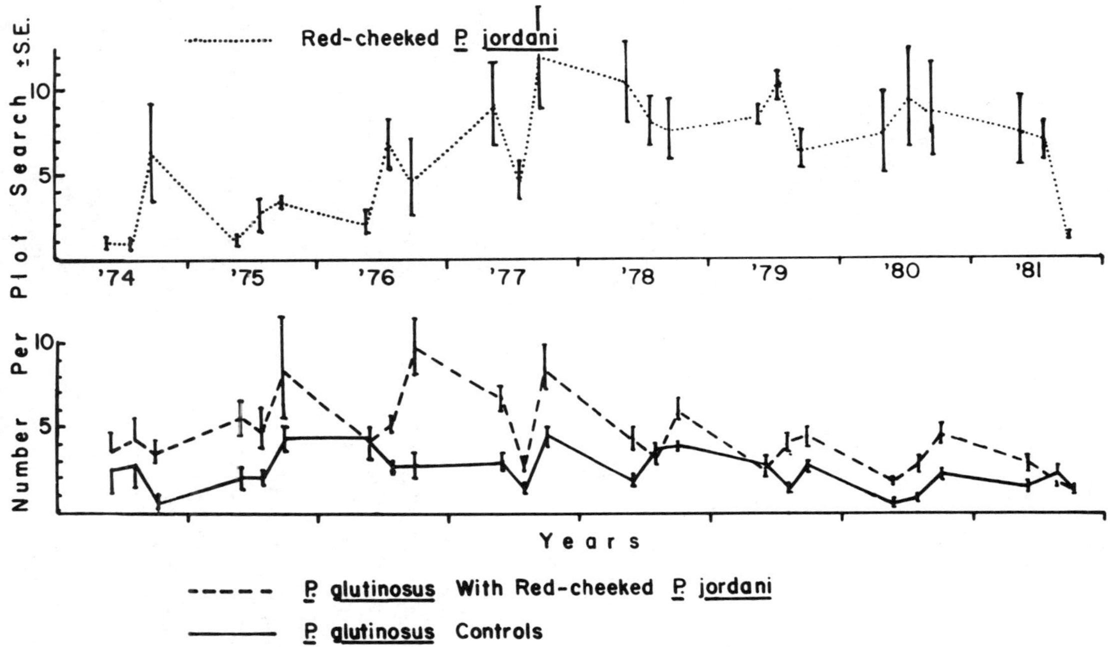

Figure 7.1. Effect of the highly competitive red-cheeked *P. jordani* on *P. glutinosus* in the Balsam Mountains, where competition beween *P. glutinosus* and the local gray-cheeked *P. jordani* is not intense. (From Hairston 1983a.)

mens on plots from which both local species had been removed fared the same as the ones represented in Figure 7.1. Thus, the Balsams *P. glutinosus* had no detectable effect on the Smokies *P. jordani*, even though the two species were shown locally in the Balsams to have mutual negative effects (see Chapter 5).

In the Great Smoky Mountains, there was no difficulty in establishing populations of *P. jordani* from the Balsams, and the experiments were terminated after five years. The results were clear, and in marked contrast to the results from the reciprocal experiment (Fig. 7.2). In the presence of the introduced gray-cheeked specimens from the area of wide altitudinal overlap, the *P. glutinosus* population increased markedly, a result very similar to that in the experiment in which the local population of *P. jordani* was simply removed. The comparison showed that *P. glutinosus* in the Smokies was virtually unaffected by *P. jordani* from the location where competition was less intense. In contrast to the reciprocal experiment, there was a striking difference in the performance of introduced *P. jordani* on plots with and without local *P. glutinosus*. They fared better where both local species had been removed, showing the negative effect on plots where *P. glutinosus* remained.

Assuming, as is done in Chapter 5, that the relationships can be described by the Lotka–Volterra competition equations, it is possible to calculate the coefficients of competition directly from changes in the experimental populations (Table 7.2). The values dramatize the respective competitive effects between the representatives of the two species in the different areas. For both species combinations, there is no effect of the Balsams population on the other species from the Smokies. In contrast, both species from the Smokies have stronger effects on the other species from the Balsams than the Balsams populations have on each other.

If the populations in the area of less competition had evolved to have different ecological requirements, they should have had some effect on the other species from the Smokies, because of the remaining within-Balsams competition. That is, there is some remaining overlap between the two species in ecological requirements, and that should be reflected in an effect on congeneric species from the Smokies. The fact that they do not have such an effect suggests that the interaction is a direct behavioral one, with the representatives from the area of more intense competition (the Smokies) having a complete advantage because they have evolved to be more aggressive toward each other, and the additional aggression favors them against their congeners from the Balsams.

That possibility was investigated by my former student, Dr. Kiisa C. Nishikawa. Her experiments are mentioned briefly in Chapter 5. She took advan-

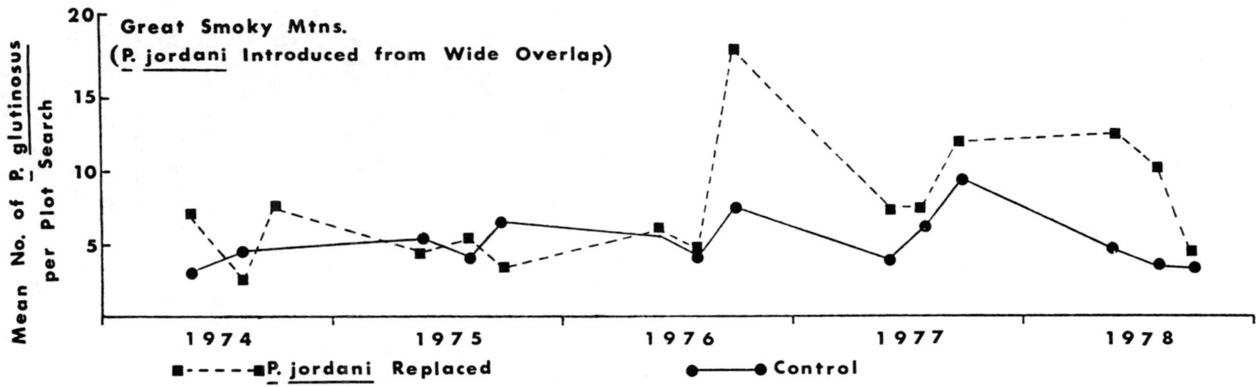

Figure 7.2. Average number of *P. glutinosus* per plot search on plots where *P. jordani* from an area of wide altitudinal overlap was being substituted for *P. jordani* in an area of narrow overlap. For the experimental data (dash line), each point represents the mean of four searches; for the controls (solid line), each point represents the mean of six searches. (From Hairston 1980c.)

Table 7.2. *Coefficients of competition between* P. jordani *and* P. glutinosus *in the Great Smoky Mountains and the Balsam Mountains, and for combinations of the two species between the two areas*

Within areas	Great Smoky Mountains	Balsam Mountains
$\alpha_{g,j}$	0.194	0.142
$\alpha_{j,g}$	2.255	0.629
Between areas	Smokies *jordani* vs. Balsams *glutinosus*	Balsams *jordani* vs. Smokies *glutinosus*
$\alpha_{g,j}$	0.403	0.029
$\alpha_{j,g}$	0	0.874

Source: Hairston (1980c).

tage of the fact that the salamanders show much of their behavioral repertoire in circular plastic containers 21 cm in diameter. When kept on moist filter paper, an individual marks it with skin secretions and feces and is then able to recognize its own filter paper as different from that marked by another individual. A specimen on its own filter paper in a plastic dish is a "resident" and behaves aggressively toward an additional specimen of the same sex that has been introduced into the dish. Nishikawa found no difference in the aggression between males as compared with that between females. She conducted two sets of experiments, comparing the aggression of residents toward intruders. The comparisons were between the behavior of individuals of each species toward members of the same species and toward members of the other species. Representatives from each mountain range were observed in both sets of experiments. In the first set (Nishikawa 1985a), the average performance of residents toward conspecific and congeneric intruders was compared for each mountain range. In this set, each resident was used only once. Among specimens from the Great Smoky Mountains, there was significantly greater aggressiveness by *P. glutinosus* toward *P. jordani* than was the case among specimens from the Balsam Mountains. This difference between specimens from the two mountain ranges was not observed intraspecifically in *P. glutinosus*. This experiment supports the hypothesis of alpha selection on *P. glutinosus* in the Smokies, the area of narrow altitudinal overlap and stronger interspecific competition. No such difference in mean aggressiveness was observed among resident specimens of *P. jordani*. Neither toward conspecific nor toward heterospecific individuals was there a significant difference in aggressiveness between specimens from the two mountain ranges. The difference between the behaviors of the species may be related to their relative abundance. *P. jordani* is six to eight times as abundant as *P. glutinosus* in the areas of common occurrence, and on the basis of chance encounters, *P. glutinosus* is far more likely to meet a heterospecific individual than is *P. jordani*. Thus, selection based on interspecific interactions is more likely to affect *P. glutinosus*. This consideration prompted the second set of experiments.

In this set, each resident individual was presented with a conspecific intruder and with a heterospecific intruder at two-day intervals, in random sequence (Nishikawa 1985b); intruders were from the same area as the residents. Rather than an average performance by different residents, these experiments tested differential performance by individuals. *P. jordani* from the Balsam Mountains were significantly more aggressive toward conspecific intruders than toward *P. glutinosus*. The result was in sharp contrast to that obtained on individuals from the Great Smoky Mountains. *P. jordani* from there were more aggressive toward *P. glutinosus* than toward conspecifics, although the difference was not statistically significant. The contrast between the specimens from the two areas shows the relative increase in aggressiveness toward heterospecifics in the Smokies and is further evidence for alpha selection. Resident *P. glutinosus* from the Smokies were more aggressive toward both conspecifics and heterospecifics than were those from the Balsams.

Thc results of the two sets of experiments on the behavior of these salamanders toward intruders confirm the conclusions reached from the experiments on transplanting populations of *P. jordani* between the two mountain ranges. In the area where interspecific competition had been found to be more intense, it was shown to be related to a higher level of aggression toward heterospecifics, relative to behavior toward conspecifics, than was the case where competition was shown to be less intense. This resulted in strong negative effects of Smokies salamanders on Balsams specimens of the opposite species, with no detectable effect vice versa. The evolution of increased aggressiveness in the Great Smoky Mountains has resulted in each species being able to exclude the other from the altitudes to which it is better adapted, giving a narrow zone in which neither can win in competition. In the Balsam Mountains, where alpha selection has not occurred, competition is not strong enough to result in the mutual exclusion seen in the Smokies (Hairston et al. 1986).

Small plethodons

Competition between *Plethodon cinereus* and *P. shenandoah*, described in Chapter 5, has not been pursued from the viewpoint of evolutionary impact, as has been the case with the large plethodons. Nevertheless, the extensive experimentation on the aggressive behavior of *P. cinereus* by Jaeger and his students has shown that it is territorial and that it defends moist refuges in dry weather. Furthermore, the two species distinguish and avoid each other's secretions, suggesting a detectable level of interspecific territoriality. The asymmetrical competition found in the field indicates that *P. cinereus* has evolved a sufficiently aggressive behavior to prevent any extension by *P. shenandoah* of its range outside the area from which *P. cinereus* is excluded physiologically. The behavior of the latter may only be incidental to its intraspecific territoriality, and only further experimentation can resolve the question.

Evolutionary ecology in streamside communities

The ecological importance of predation in a community of *Desmognathus* is described in Chapter 6. If we can make the assumption that the present ecological relationships provide clues to the pressures of natural selection that have prevailed in the past, *Desmognathus* is a key genus in understanding the evolutionary history of the Family Plethodontidae. This is the most abundant, widespread, and ecologically varied family of salamanders. It contains more species than all of the other seven families combined (Table 3.1). Its geographic distribution and the life histories of its species have led herpetologists to conclude that the family arose in Appalachia. It is reasonable to assume that the original life history was an amphibious one, and all species of Plethodontidae having a larval stage are confined to eastern North America. The most remarkable trait shared by all species is the absence of lungs. In some other salamander families, reduction of lungs has occurred in some species, and this is always associated with the mountain-stream habitat. Lunglessness would be an advantage in mountain streams, because lungs act as floats, and floating might be a decided disadvantage when it could result in being dashed against rocks. The skins of many amphibians are respiratory organs, and in the highly oxygenated water of mountain streams, the skin would provide for adequate respiration. Loss of lungs would therefore be no disadvantage.

Appalachia was uplifted at the end of the Paleozoic Era and has remained so ever since. Thus, the appropriate habitat for the origin of the Plethodontidae has been available for a sufficient time, and its central position within the range of all species having the probable original life history is convincing evidence for locating the place of origin. Only the two tribes that are independent of water for a larval stage have spread beyond the eastern part of North America. These two tribes are the Plethodontini and the Bolitoglossini. Representatives of the former tribe are found in both eastern and western North America; the Bolitoglossini are represented in western North America, Central America, South America, and Europe. As far as can be determined, the terrestrial way of life arose independently in the two tribes. Because all species in both tribes lack a larval stage, they are valueless in attempts to discover the ecological conditions that led to their becoming completely terrestrial.

The Subfamily Desmognathinae, however, contains species that span the entire range of habitats, from one species that remains aquatic throughout its life, despite undergoing a complete metamorphosis, through all intermediate degrees of partially aquatic and partially terrestrial existence, to two species that have dispensed with the larval stage. This is not the place to discuss the science of cladistics, which is concerned with establishing the evolutionary relationships among a group of species, but by proper application of the principles, decisions can be reached with some assurance as to which one of the group is most like the common ancestor. In the case of *Desmognathus*, there is agreement that one of the large streambank species, either *D. quadramaculatus* or *D. monticola*, occupies that position in the genus. That being the case, the ecological progression toward a terrestrial existence can be taken as a model of the evolution of the genus, with *D. wrighti* and *D. aeneus* representing the end point. The ecological relationships among these species, then, should reveal the selective pressures that have led to terrestrialism and, in the absence of any evidence internal to the two terrestrial tribes, could indicate the most important ecological factors in the evolutionary history of the whole family Plethodontidae. The problem can be stated as follows: Having evolved lunglessness as an adaptation to mountain streams, what was the selective pressure that led to their abandoning the ancestral habitat?

The effect of each species of *Desmognathus* on the others in the community is described in Chapter 6. These ecological relationships were determined by experimental removals of each of the two species in intermediate habitats over a period of four years and by other experiments on mutual behavioral interactions under a variety of specific conditions. The most direct effects involve aggression, both between and within species, with the larger individual always being the aggressor. When there is a sufficient difference in sizes, the aggression becomes either predation or cannibalism. The removal experiments established that the relationship between *D. ochrophaeus* and *D. quadramaculatus* (the largest and most aquatic species) and between *D. ochrophaeus* and *D. monticola* (the next largest) was predation on *D. ochrophaeus*. The evidence was that when the prey species was removed from plots, the populations of both of the larger species declined relative to their populations on control plots. If competition had been of any importance, the populations of the larger species should have increased. Competition was demonstrated in the case of the relationship between the two large species. The removal of *D. monticola*, the smaller, eventually resulted in an increase in the abundance of *D. quadramaculatus*, a result not predictable from a hypothesis of predation. The larger *D. quadramaculatus* has been observed to capture and eat *D. monticola*, but no experiment was conducted to discover if such an event is sufficiently frequent to have an impact on the population of *D. monticola*. From these experiments it is clear that both predation and competition occur within the community, with predation being the more important interaction.

There is a good case to be made for predation in and beside the streams being the selective pressure that repeatedly caused different groups of Plethodontidae to become terrestrial. Competition could have been involved in a certain amount of niche partitioning, such as one species becoming completely aquatic, but the reversal of expected size relationships among the species of *Desmognathus* argues strongly against competition being the important selective influence.

These readily demonstrable interactions among three species make the relationships of the fourth species both puzzling and fascinating. Removal of either *D. ochrophaeus* or *D. monticola* had no effect on the population of *D. aeneus*, the smallest and most terrestrial member of the community. In its behavior, *D. aeneus* is unlike the other species. It is frequently found climbing on shrubs and other low plants during especially humid nights, but I have never observed *D. aeneus* above the forest floor. In fact, it is rarely seen even on the surface. All other salamander species with which I am familiar are more readily found in the open at night than they are by searching under cover objects during the day. For example, during four months early in the removal experiments, night and day searches were conducted on the same nine plots. *D. quadramaculatus* was seen 3.75 times as frequently at night; *D. monticola* was seen 1.81 times as frequently; and *D. ochrophaeus* was seen 1.65 times as frequently. In contrast, *D. aeneus* was seen only one-sixth as frequently (Table 7.3). This peculiarity is not a result of its small size, because *D. wrighti*, which is equally small, climbs readily (Hairston 1949). These observations on *D. aeneus* were even more surprising when I found the following account in E. O. Wilson's book *Biophilia*. He went "to a woodlot in central

Table 7.3. *Numbers of* Desmognathus aeneus *seen on night and day searches on the same nine plots at Coweeta Hydrologic Laboratory*

	Total no. seen	
Plot no.	4 night searches	4 day searches
1	0	0
2	0	2
3	1	6
4	0	2
5	1	1
6	1	2
7	0	3
8	0	0
9	0	2

Alabama late at night in a cold rain to look for frogs and salamanders. I had been to the site several times before on sunny days and seen nothing. That night, as soon as we walked into the woods, we found a teeming population of a pygmy salamander of the genus *Desmognathus*, recently described by zoologists as a new species. The delicately built amphibians, which resembled shiny popeyed lizards, were climbing up grass and low bushes. They jumped agilely from branch to branch in search of prey" (Wilson 1984, p. 104). The salamanders that Wilson saw were *Desmognathus chermocki*, a name since combined (by other zoologists) with *D. aeneus*. The difference in behavior between members of the same species in central Alabama and southwestern North Carolina is striking and demands an explanation. The most obvious difference, from the standpoint of effect on behavior, is the absence in central Alabama of the three congeneric species of *Desmognathus* with which *D. aeneus* is associated in North Carolina. The population in central Alabama is isolated from the rest of the species and presumably has not been subjected to selective pressures that would cause it to change from the usual behavior of terrestrial salamanders. The hypothesis to explain the situation is that under the pressure of large predatory congeners, *D. aeneus* has become atypically secretive, so much so that those species no longer have a detectable effect on its population. In Chapter 5, mention is made of the apt phrase "the ghost of competition past" to describe evolution of differences as a result of an original competitive situation that is no longer detectable. With *D. aeneus* in the Nantahala Mountains of North Carolina, we may be seeing "the ghost of predation past."

Aquatic life and loss of the adult body form

The most ubiquitous trend in altered life histories has been the tendency for some species of salamanders to become neotenic. At least a few species of all families have done so, and for three families, no other life history is known (Table 3.1). Ecological experiments of the kinds described earlier in this chapter have not been performed on neotenic species, and the ecological pressures involved are mentioned here for the sake of symmetry, rather than implying the assurance that experiments provide.

Three hypotheses have been advanced to explain the advantage of losing the adult form. The first of these is the harshness of the local terrestrial habitat. This explanation has been advanced for populations of *Ambystoma tigrinum* in western North America, the Mexican axolotls, and the populations of *Notophthalmus viridescens* on the sandy coastal plain of New England. A second group of neotenes is composed of some members of the plethodontid Tribe Hemidactylini, which live in or close to subterranean habitats. Again, the adult habitat, though not harsh in the physical sense, is frequently devoid of food. The European *Proteus anguinus* is also subterranean. Neither hypothesis will suffice for the large species of three families in the eastern United States: the Sirenidae, the Amphiumidae, and the proteid genus *Necturus*. All inhabit streams and ditches in humid areas where there are many species of terrestrial salamanders, and the physical environment cannot account for the loss of the adult stage. It has been suggested that large terrestrial adults attract large mammal predators, but the hypothesis appears to be untestable and sterile.

It appears that the origins of terrestrialism are quite different from the origin of neoteny, and community ecology did not play a part in the latter.

Synthesis

The charge that natural selection is a tautological argument has been made repeatedly throughout the history of evolutionary thought. The claim has generated vigorous debate, and like many other philosophical discussions, it is rarely resolvable. The possibility that any part of selection theory can be advanced experimentally should be welcome, and this chapter reports two examples of testable hypotheses about the evolution of salamanders. Both tests require the assumption that present ecological and behavioral phenomena represent a continuation of past selective pressures. Nevertheless, any experimental test of an evolutionary hypothesis is welcome.

The first series of tests was designed to permit a choice to be made between two contrasting hypotheses about the evolution of competing species. One is the dominant opinion that competing species will evolve to become different in their use of a resource for which they are competing, thus permitting their coexistence. The second hypothesis is that one or both species should evolve to become better competitors. This hypothetical phenomenon could lead to their being respectively dominant in different habitats, or different parts of an ecological gradient, and therefore to "partitioning" the habitat in some sense. However, the process by which the result is obtained would be very distinct from conventional niche partitioning.

The species involved are the terrestrial *Plethodon jordani* and *P. glutinosus* of the southern Appalachians. They had been shown to compete strongly in some areas, and weakly in another area. There is good evidence that they do not compete for food and that this is not a result of partitioning of that resource in the area of weak competition. Reciprocal transfer of *P. jordani* between areas established the competitive superiority of both species from the area of intense competition over their congeners from the area of weak competition. Neither of the latter species had any effect on the strongly competing ones, even though they did compete with each other. If they had evolved a somewhat different resource use, there should have been some effect on the species from the area of stronger competition, and their failure to do so suggested that a behavioral dominance had evolved in that area. A series of experiments established that *P. glutinosus* from the area of intense competition is more aggressive toward *P. jordani* than are individuals from the area of weak competition. It was also found that *P. jordani* was significantly more aggressive intraspecifically than interspecifically in the area of weak competition, a difference not found in the area of strong competition. The difference between the species is best understood when their relative abundances are taken into account. *P. jordani* is six to eight times as abundant as *P. glutinosus*, and the latter is much more likely to encounter a heterospecific individual. Taken together, these experiments confirm the hypothesis that extra aggression is the basis of the finding that competition is stronger in one area than in the other and strengthen the hypothesis that natural selection has been for this competitive ability.

The second situation for which experimental tests were devised concerns the evolutionary history of the genus *Desmognathus* and to some extent the evolution of the Family Plethodontidae, which are the most abundant, widespread, and varied salamanders. *Desmognathus*, in combination with the closely related *Leurognathus*, spans the entire set of habitats from never leaving the water to never entering it. The combinations that occur together seem to conform to the general observation of such groups, in that they occupy distinct though overlapping distributions and appear to partition the habitat into aquatic, streambank, and terrestrial zones. Their food reflects their different habitats, and for a time they were considered to be a typical example of competing related species that had evolved sufficient differences to permit coexistence. Eventually it was pointed out that there is an inconsistency that the smallest and most recently evolved species are the most terrestrial. Because these species lack lungs and must respire through the skin, water must be lost through the moist skin, and this would be relatively greatest in small specimens, in contradiction to the principle that niche partitioning should increase efficiency. The countering hypothesis was proposed that predation by the larger, more aquatic species was the selective pressure involved in the evolution of the genus. Removal of a smaller species should favor a larger one only if competition is important; if the larger species is a predator on the smaller, it should not gain and might lose by the removal of a prey species. When a small species was removed, both of two larger ones decreased in numbers, showing that the small one is an important prey item for both. As suggested by that result, the two larger species are competitors, demonstrated to be so by the fact that removal of the smaller of the two resulted in an increase in the abundance of the larger. A fourth species, smallest and most terrestrial of the community, was not affected by the removal of either of the next two larger species. It has apparently evolved to become unusually secretive, which has not happened in an isolated part of its range, where its larger, predatory congeners are absent. The *Desmognathus* community was shown by these experiments to fail to bear out predictions based on the dominant theory of community evolution.

That theory has been derived by producing plausible explanations for available observations, frequently involving mathematical manipulations. Such an approach avoids the strong inference of experimental science, and the assumptions are often unreasonable when stated in terms that are not mathematical. The theory has failed both of two experimental tests involving the evolution of salamanders.

8

Salamanders in future ecological research

Introduction

Personal acquaintances among a limited number of ecologists, plus individual preferences for kinds of animals, have resulted in a disproportionate amount of community theory being based on knowledge of birds and lizards. The rich literature on marine communities has contributed relatively little to mathematical theory, an exception being the work of Paine and Levin (1981). These more-or-less fortuitous circumstances have produced a bias away from consideration of experimental results and toward "natural experiments" as interpreted from bird and lizard ecology. Birds are notorious for their intractability to field experiments (Wiens, Rotenberry, & Van Horne 1986; Wiens 1986), and experiments with lizards have given results as inconsistent as those with different groups of salamanders (see Chapter 1).

The combination of a restricted base and a bias away from field experiments has tended to cause theoreticians to ignore information that was not consistent with existing theory, or to attempt to force consistency when it was not obvious. Thus, Schoener (1983) argued that my failure to obtain an effect on *Plethodon serratus* and other plethodontids from the removal of either *P. jordani* or *P. glutinosus* (Hairston 1981) was due to the shape of the zero isoclines. Had they followed his model (Schoener 1978), a more complete removal of *P. jordani* than I obtained would have been required to affect the other species. The natural history of *Plethodon* and the requirements for consistency with the model are strongly against such a scenario (Hairston 1985).

It is my hope and expectation that the preceding chapters will have two effects: to call the attention of ecologists to the need for improvements in community theory based on the variety of results that have been obtained from the large number of field experiments, and to show the superiority of salamanders as material on which to conduct further experiments. Many of the opportunities for community ecology that are provided by salamanders remain unexplored. Some of these are described in this chapter.

Possibilities from the variety of salamander communities

Salamander communities in ponds

Community theory predicts that addition or subtraction of species will alter the equilibrium density of the other species present. The prediction has been verified in experimental manipulations of pond-inhabiting communities, although the interaction has frequently been found to be one of predation within the community, rather than the predicted competition (see Chapter 4). Two genera seem unusually well suited to further investigation of salamander communities in ponds: *Triturus* and *Ambystoma*.

For the former, Avery (1968), Szymura (1974), and Griffiths (1986) have published studies of the comparative ecology of three different combinations of species, and they differed on their findings on the expected differences (see Chapter 4). The migratory habits of these species suggest that differential interception of different species should permit the testing of hypotheses about competition among them. Moreover, community ecology makes predictions about niche breadths for different numbers of species. The combination of such natural experiments with manipulations should eventually make this genus irresistible to British and European ecologists.

Experiments on different combinations of species of *Ambystoma* are described in Chapter 4. The possibilities are by no means exhausted. Six species are known to occur on the coastal plain of South Carolina, and the work of Semlitsch has provided an excellent background on some of them for further community experiments. The more that is known about any members of a community, the more precise predictions can be about community interactions among them. Similarly, Wilbur's papers on the communities in Michigan can be used in generating predictions about those species, only two of which are in common with the South Carolina fauna.

Terrestrial salamanders

Among purely terrestrial species, the interaction has been competition, no predation having been detected experimentally, although there have been anecdotes of individual examples of predation. Experimental manipulation of *Plethodon*, both the *P. jordani–P. glutinosus* and the *P. cinereus–P. shenandoah* pairs, has revealed the expected competition between the species (see Chapter 5). The experiments have also shown that in this genus, only members of the same species group have demonstrable effects on each other. The members of each of the two groups mentioned have been found to be very similar from the standpoint of enzyme polymorphisms, but there are great differences between one group and the other (Highton & Larson 1979). The experimental evidence shows that ecological requirements must be alike within a group and sufficiently different between groups to permit coexistence without competition. Highton and Larson recognize eight species groups within the genus. Not all of these provide sympatric or even parapatric examples, but enough of them do to be useful in testing predictions from natural experiments as well as from further manipulative ones. For example, in the southern Appalachians, the French Broad River limits the westward distribution of *Plethodon yonahlossee*. East of the river, there are three species of the *glutinosus* group: *P. glutinosus*, *P. jordani*, and *P. yonahlossee*. The experiments reported in Chapter 5 were carried out west of the river, in the absence of *P. yonahlossee*. Competition was found in different degrees in two parts of the area. What is the predicted effect of the third species on the competitive relationships between *P. glutinosus* and *P. jordani*? It is known that in the Black Mountains, these two species have altitudinal distributions similar to those where competition is intense (Hairston 1949, 1980a). This suggests that the requirements of *P. yonahlossee* do not lie between those of the others, and its removal from experimental plots should have a significantly greater effect on one of them than on the other. Its altitudinal distribution overlaps more with *P. glutinosus* (Hairston 1949). Reasoning from the experiments described in Chapter 5, competition should be almost as intense between *P. jordani* and *P. yonahlossee* as it is between *P. jordani* and *P. glutinosus*. Here are two predictions for future experimentally inclined community ecologists:

1. In the areas of altitudinal overlap, the numerical abundance of *P. jordani* should be less in the Black Mountains than in either the Balsams or the Smokies.
2. Removal of *P. yonahlossee* should have a proportionally greater effect on *P. jordani* than on *P. glutinosus*.

The foregoing discussion implies that genetic relatedness can be used to predict the intensity of competition, at least in the comparative sense within a taxonomic group. This has been an implied assumption in many studies of community ecology, and the extent of its applicability needs to be tested in a rigorous manner. The phylogenetic tree presented by Highton and Larson could be used to generate predictions, such as which pairs of species should be expected to compete most strongly where three or more congeneric ones occur together.

In addition to the members of the genus *Plethodon* in the eastern United States, promising terrestrial communities are those mentioned in Chapter 5. Lynch's data (1985) on the food of the species in northern California provide a good background for experimental investigations on the communities involved. Finally, the tropical plethodontids occur in many combinations, and with the new knowledge of the taxonomic relationships, as shown by biochemical and immunological work (Larson 1984), predictions based on community theory should be precise enough to be testable. The predictions will be made with more confidence among terrestrial salamanders than in pond-dwelling or streamside communities because ecological interactions in the other two have been shown to be confounded among competition and predation.

Communities of streamside salamanders

Nevertheless, the Subfamily Desmognathinae is tempting because of the variety of combinations of coexisting species. Within the southern Appalachians, at least twelve different combinations of coexisting species are known. Although several of them have been studied from the standpoint of behavioral interactions that are highly relevant to competition and predation, only one has been subjected to long-term manipulative experiments to reveal populational responses. The results that have been obtained in the experiments suggest predictions about the interactions that are to be expected in other combinations of species. Combinations that are predicted to involve predation are *D. quadramaculatus* with almost any other species, and *D. monticola* with *D. imitator*, *D. fuscus*, or *D. santeetlah*, and those predicted to involve competition are the *Leurognathus–D. quadramaculatus*, the *D. ochrophaeus–D. imitator*, the *D. ochrophaeus–D. fuscus*, and the *D. imitator–D. santeetlah* pairs. Some of these predictions would be mutually invalidating. For example, the presence of the large predator *D. quadramaculatus* might well have a sufficient effect on both *D. ochrophaeus* and *D. imitator* to negate competition between them where the three species occur together in the Great Smoky Mountains.

In seepages and streams where *D. quadramaculatus* does not occur, community theory makes predictions about the level of competition between *D. ochrophaeus* and *D. santeetlah*. They should compete more in the Balsam Mountains than in the Great Smoky Mountains, because in the latter location the similar-sized *D. imitator* occupies a habitat that is intermediate between those of the other two (Tilley et al. 1978; Tilley 1981). Experiments with these three species would yield information about the community relationships, if any, of *D. wrighti*. The comparison between that species and *D. aeneus* (see Chapter 6) would be valuable, because they are the only completely terrestrial members of the subfamily. Both species are known to occur in the Nantahala

Mountains, but *D. wrighti* is apparently too uncommon there for quantitative investigations to be feasible.

The foregoing predictions are only examples of the opportunities for future research on the community ecology of Desmognathinae in the southern Appalachians. The work described in Chapter 6 has established that much can be learned from short experiments in pens, because aggressive behavior plays such a large part in the ecology of these salamanders. As discussed in that chapter, however, such behavior can involve defense of either a refuge or an area or can involve predation. The resolution of the difference may require long-term experiments that reveal populational responses to manipulations. The possibilities in this field are limited only by the ingenuity of community ecologists and the time that they are willing to invest in the research. The field is open.

Behavioral ecology and community studies

Salamanders require only a little space in which to carry out their behavioral activities. This happy circumstance has been known for more than fifty years, starting with the work of Noble and Brady (1930). Both courtship and aggression can be observed in circular containers 20 to 30 cm in diameter (Organ 1958, 1960; Arnold 1976; Nishikawa 1985a,b), and except for the inability of losers in aggressive encounters to escape, the observations agree with those made in the field under much more difficult conditions (Arnold 1976). The detailed studies of the community ecology of salamanders in all kinds of habitats, as described in Chapters 4, 5, 6, and 7, have the common finding that all of the ecological interactions, whether competitive or predatory, involve aggression.

Competition and aggression

In situations involving interspecific competition, the terrestrial genus *Plethodon* has been the object of most of the experimental work. Jaeger and his students (Jaeger 1979, 1980, 1981; Jaeger & Gergits 1979; Jaeger et al. 1983) have established that *P. cinereus* defends territories by aggressive behavior and that some of this is directed toward the congeneric species *P. shenandoah*. Nishikawa (1985a,b) has shown how the relative amount of interspecific aggression between *P. glutinosus* and *P. jordani* determines the amount of distributional overlap between these species on different mountain ranges. In both examples, the composition of the salamander communities depends on their behavior. These pioneering experiments point the way to an understanding of other communities of terrestrial salamanders, most of which remain to be investigated. The communities in northern California and in the American tropics include intergeneric combinations of potentially competing species. It would be worthwhile to search for aggressive interactions among them, especially when niche or distributional relationships suggest ecological impact.

It is also possible that observations of behavior could reveal more subtle relationships than have been discovered through manipulative experiments. Such observations involve much less in the way of long-term commitments than do removals and introductions, and they would permit tests of possible interactions between additional species beyond those already known to influence each other ecologically. For example, I found no effect on *Plethodon serratus* when I removed either *P. jordani* or *P. glutinosus*. Would there be no behavioral interaction under the laboratory conditions used by Jaeger and by Nishikawa? The known importance of aggression suggests that such experimental tests could resolve the question, raised by Schoener (1983), of whether or not my removal experiments were sensitive enough to reveal interactions relatively weaker than the ones that I found.

Much of community theory rests on the assumption of linear arrays of resource states. The importance of aggression in communities of salamanders suggests that defendable resources are the ones that are relevant to the species involved. This raises the question of what kinds of resources can be both defended and arrayed in a linear manner. Space is the most obviously defendable resource, and its partitioning intraspecifically through territoriality has been documented in two cases. It is difficult to visualize how space can be partitioned in the linear manner depicted in community theory. If our present understanding of salamander communities stands up under further investigation, modification of that theory, or its restriction to some specified communities, will be required.

Predation and aggression

A mixture of competition and predation has been found to characterize communities of salamanders in ponds and at streamsides. Which is more important in different combinations seems to depend on the relative sizes of the individuals involved (Stenhouse et al. 1983). This relationship can be quantified in a precise way by laboratory experiments. The desirability of knowing the ratio of sizes comes from the fact that whereas neither of two competitors gains from the interaction, a predator would derive benefit. The size distributions within the populations could then be used to estimate the impact on both coexisting species. One would suppose that head widths would provide the best indicators of relative size, at least for the critical size ratio at which competition becomes predation, but snout–vent lengths or weights might be important.

In addition to determining the sizes at which competitive aggression becomes predation, both intraspecifically and interspecifically, these experiments could determine the effect, if any, of prior occupancy of an experimental container. There are attractive possibilities in these proposed experiments, especially for quantifying the interactions among the species of *Desmognathus*.

In communities of *Ambystoma* species, the important determinants of competition versus predation are timing of breeding and relative growth rates. The advantage of early breeding, and hence attainment of a size large enough to be the predator, is obvious for *A. opacum*, whereas *A. tigrinum* may gain more from rapid growth than from slightly earlier breeding. These possibilities could be explored by straightforward experiments, because it is easy to control the rate of growth of early-breeding forms by controlling the temperature at which they are kept until later-breeding forms appear.

Mating behavior and community studies

Interest in the mating behavior of salamanders was stimulated by the hope that the details would be helpful in understanding the phylogeny of the group. It is only recently that reproductive isolation has been tested through studies of mating behavior. In community studies, this information will be useful where there is a question of whether two forms have a parapatric distribution or a zone of secondary hybridization. In the former case, as in the narrow altitudinal zone where *Plethodon glutinosus* overlaps with *P. jordani*, the community contains one more species than is the case where there is free intergradation, as is true of the same two species in the Nantahala Mountains. The addition or subtraction of a species theoretically affects the abundances of the rest of the community, and it is important to know the facts. Even where there are some intermediates, the ease with which the occasional intermediate forms mate with the others is valuable information.

Importance of salamanders in the dynamics of ecosystems

Terrestrial species

Until recently, the abundance of salamanders attracted little attention. The first surprising density estimate was that of Merchant (1972). He reported that there were 200 *Plethodon jordani* per 2,500 ft^2 (232.25 m^2), and

54 *P. glutinosus* in the same area. The mean density of large plethodons was thus estimated to be greater than one per square meter in the narrow zone of overlap in the Great Smoky Mountains. My own estimates, based on counts over longer periods in a nearby area in the Smokies and in the Balsam Mountains, were less by a factor of about 2.6, and although my data included some counts that were too low because of cold weather (Hairston 1983b), there were not enough such counts to account for the difference. The basis of my estimates is given in Chapter 2. It seems likely that Merchant's figures are somewhat inflated because of the effect of using the capture–recapture method where many individuals were not recaptured because they were on the border of his plot. Making some adjustment for the differences, it seems likely that where the two species overlap in distribution, the number of large plethodons is approximately 0.7 m^{-2}. I have made complete collections on measured plots and weighed the totals. The mean for all *P. jordani* is 1.0 g, and for all *P. glutinosus* is 1.5 g. Thus, the biomass would be about 0.83 g m^{-2}, with a range of estimates from 0.39 to 1.25. Except for very local and numerically trivial species, *Plethodon serratus* is the only other member of the genus west of the French Broad River in the southern Appalachians. It is small and much less abundant than its larger congeners, and its contribution to the total biomass is less than 0.002 g m^{-2}. The very similar species *P. cinereus* reaches much greater densities in the mountains of New Hampshire and Virginia. Burton and Likens (1975a) estimated that the Hubbard Brook ecosystem contained an average of 2,583 specimens per hectare, or 0.26 m^{-2}. In Virginia, Jaeger (1979) made repeated counts over two periods and obtained means of 2.1 and 2.3 specimens per square meter. The biomass would be comparable to the higher values reported earlier for large plethodons.

Amphibious species

In the southern Appalachians, the members of the genus *Desmognathus* make up the great majority of salamanders that are at least partly bound to the vicinity of water. Numerically, *D. ochrophaeus* is most important. I estimate that on my experimental plots, there are sixteen individuals per linear meter of stream, not including larvae. From the weights and relative abundances of the different size classes, it is possible to calculate the biomass. Within the 1,626 ha of the Coweeta Hydrologic Laboratory, there are 58.2 km of streams, or 35.79 m of stream per hectare. On an area basis, *D. ochrophaeus* accounts for 0.052 g m^{-2}. Of the two larger species, *D. monticola* contributes 0.037 g m^{-2}, and *D. quadramaculatus* 0.140. Ignoring the tiny and uncommon *D. aeneus*, the *Desmognathus* biomass averages 0.229 g

m^{-2}. The remaining salamander species in the area are unimportant from the standpoint of biomass of transformed individuals, amounting to little more than 0.01 g m^{-2}. They include three species of *Eurycea*, *Gyrinophilus porphyriticus*, *Pseudotriton ruber*, *Notophthalmus viridescens*, and several rare species.

Aquatic forms

The larvae of the amphibious species contain a negligible part of the total salamander biomass. *Eurycea bislineata* is numerically abundant, but the larvae are small, accounting for only an estimated 0.002 g m^{-2} for the whole area. The remaining species, though larger, are too uncommon to be of any importance.

There is one very large aquatic species, *Cryptobranchus alleganiensis*, for which no local abundance estimate is available. It inhabits only large streams, but individuals may weigh as much as 500 g. They could increase the total salamander biomass by a significant amount, but their impact on the forest ecosystem is probably not important, and it is the forest that is of interest here.

Total salamander biomass in Appalachian forests

From the foregiong estimates and calculations, it can be concluded that an average of 0.6 to 1.0 g m^{-2} is a conservative estimate for salamander biomass. Burton and Likens (1975a) concluded that at Hubbard Brook in New Hampshire, where the salamander biomass was only 0.177 g m^{-2}, it was equal to that for mammals and twice that for birds at the height of the breeding season. If we use their estimates for birds and mammals, farther south the salamanders easily outweigh both of them combined. For animals so high in the food chain, they have tied up an impressive amount of energy. Burton and Likens (1975b) obtained the following values in kilocalories per gram of dry weight: *Plethodon cinereus*, 5.176; *Desmognathus fuscus*, 5.051; *Eurycea bislineata*, 5.248; *Gyrinophilus porphyriticus*, 4.950. Using a rounded value of 5 kcal per gram for the species in the southern Appalachians, and their conversion factor from wet to dry weight of 0.233, I make the conservative estimate that salamanders contain an average of 1.165 kcal m^{-2} of forest. This not only exceeds the calculated value for all other predators combined but also is greater than the caloric content of most herbivores (e.g., Golley 1968).

It is interesting to speculate on the significance of having such a large amount of energy tied up in a part of the food web that itself turns over so slowly. Burton and Likens (1975b) commented on the slow metabolism of salamanders, making them less effective in the flow of energy through the ecosystem than the homeothermic vertebrates, despite containing a larger store of energy. They suggested that salamanders are a potential source of high-quality protein. However, except for *Desmognathus*, most salamanders are distasteful to predators, and predation on such forms as *Plethodon* does not appear to be an important source of mortality.

Such a large part of the energy captured in photosynthesis never enters the animal community that animal ecologists often disregard the energetic impact of animals and concentrate on population phenomena. If we consider that a large part of the energy that reaches the secondary consumers is in salamander biomass, the impact of this "storage" is worth considering. It is possible that the effect is of damping stochastically induced fluctuations in the rate of energy flow. There is an opportunity to discover details of the operation of the ecosystem in manipulating the salamander biomass over a large enough area to measure the effect. What would be the effect of persistent removal from one area, and the addition of the removed animals, alive in one case, or dead in another, to fluctuations in the rate of energy flow? The answers could reveal much about the ecosystem.

The impact of salamanders on the forest-floor part of the ecosystem must be appreciable. Burton and Likens (1975b) examined the gut contents of *P. cinereus* and calculated the rate at which that species consumed soil invertebrates. Their best estimate of an energy-flow diagram indicates that during a year, the total consumption in kilocalories amounts to five times the standing-crop kilocalories of the salamanders. They gave no data on the standing crop of energy fixed in soil invertebrates, but such an estimate has been made for hardwood forests such as that at the Coweeta Hydrologic Laboratory. D. A. Crossley (pers. commun.) gives a figure of 1 g of biomass, or 5.04 kcal m^{-2}. Feeding at the rate found by Burton and Likens, the salamanders at Coweeta would require more than one complete turnover of the soil fauna each year, because they would consume 5.8 kcal m^{-2}.

Not all of the food belongs to the soil fauna proper, however. Burton (1976) listed the different food items of *P. cinereus*, their weight, and their relative contributions to the total. At least 32 percent came from the above-ground part of the ecosystem and included caterpillars, leafhoppers, and other herbivores, as well as some predators, such as carabid and coccinellid beetles. Nevertheless, the impact of salamanders on the soil fauna should be taken seriously. The soil fauna depends on the decomposition of dead litter, as few members can accomplish that themselves. The true decomposers are the bacteria and fungi, and nearly all nonpredatory soil animals obtain their nourishment from the microflora. Part of this interaction is mutualistic, the chewing of dead plant matter resulting in improved surfaces for the bacteria and fungi, but consumption of the latter appears to be an important source of mortality and possibly a cause of a reduction in the rate of decomposition.

The salamanders, in having an important impact on the animals feeding on the microflora, might be promoting a more rapid rate of decomposition and thus making possible a faster recycling of mineral nutrients in the ecosystem. This potential aspect of salamander ecology has not been considered by past investigators. It would almost certainly reward an intensive research program.

The future of community ecology

Present community theory is unacceptable to a large number of ecologists, because for it to remain internally consistent, too many facts must be ignored. These facts come from two sources: the large number of ecological experiments, and the observations on some communities that reveal the importance of stochastic fluctuations. The former, while confirming that interspecific competition is common in many situations, as expected in community theory, also show that predation is prevalent in some communities, in addition to revealing that the occurrence of competition frequently is not found when it is expected. The effects of stochastic factors, bringing about the "density-vague" effect of Strong (1984), are to make predictions imprecise.

If community ecology is to progress from its present unsatisfactory state, its conceptual base should be broadened. Attempts to incorporate predation, in the form of food-web theory, represent a step in the right direction, but such theories have thus far not been able to include competition at the lowest trophic level.

In a different conceptual mode, the idea that different trophic levels are characterized by different mechanisms of population regulation has been in the literature for more than a quarter of a century (Hairston et al. 1960), but only recently have the implications been recognized by community theorists (Connell 1983; Schoener 1983), and theoretical development is still lacking. The idea is supported by enough data and experimental evidence to be taken seriously, despite attacks by advocates of the universality of competition, the importance of plant defenses, the importance of density-independent forces, and the belief in group selection for ecosystem properties. It deserves serious consideration by open-minded theoreticians, as it could lead to noncircular predictions that have so far eluded pure competition-based theory.

Research on salamander guilds has produced some patterns of interactions that might be helpful in constructing a more general theory than is presently available. There has been no evidence of any interaction other than competition among terrestrial salamanders, and indeed there is no ecological evidence that terrestrial salamander populations are affected by predation from any source. Yet predation is common among pond-dwelling and streamside salamanders. The similarity goes no further, however. In ponds, the interactions are among the larvae, and no evidence has been found for either competition or predation among the adults. In contrast, the adults of streamside species are known to be both predators and competitors. Thus, it does not appear to be the existence of a larval stage that is related to the kinds of interactions among species with an aquatic phase that contrast with the interactions among terrestrial species. Although no explanation is obvious, these patterns may eventually be worked into a reasonable theory. The role of salamanders in terrestrial systems is consistent with the idea, expressed earlier,that different trophic levels are characterized by different mechanisms of population regulation. Because the litter does not accumulate, the decomposers must eventually consume all of it, and in that case there should be competition among the bacteria and fungi, which are the true decomposers. That would be difficult if the animals consuming these decomposers were not held in check by some means other than competition among themselves. Salamanders play an obvious role as potential regulators of the populations of the so-called saprovores and thus might be expected to compete among themselves. It is possible that analogous reasoning can be applied to pond and stream communities, but with different expectations. In ponds, debris does accumulate, with ultimate filling of the pond; in streams, there is no opportunity for debris to accumulate, as fragments are washed away. Thus, the initial observations in the three situations are quite different, and the reasoning following from those observations might well lead to promising hypotheses about the functioning of the respective salamander communities.

Finally, there are the puzzling examples in which no effect of some species on others in the guild could be shown, even by careful experimentation. At least one of these examples provides an opportunity to test an aspect of community theory. If the failure to obtain a positive result is due to a type of competitive interaction that could be demonstrated only by a more complete removal than was achieved, the hypothesis could be tested by more intensive effort (Hairston 1981, 1985; Schoener 1983). The existence of an interaction could also be sought in behavioral responses in the laboratory, because the species involved, *Plethodon jordani* and *P. serratus*, are known to be aggressive under such conditions.

Despite its obvious shortcomings, theoretical community ecology may have the potential to be really predictive. We can uncover that potential only by considering the full range of experimental results on natural communities, and salamander communities offer excellent opportunities for testing old predictions, as well as those that remain to be proposed.

References

Abrams, P. 1975. Limiting similarity and the form of the competition coefficients. *Theoretical Population Biology* 8:356–75.

Adams, J. 1985. The definition and interpretation of guild structure in ecological communities. *Journal of Animal Ecology* 54:43–59.

Allee, W. C., A. E. Emerson, O. Park, T. Park, & K. P. Schmidt. 1949. *Principles of Animal Ecology*. Philadelphia, W. B. Saunders.

Anderson, J. D. 1968. A comparison of the food habits of *Ambystoma macrodactylum sigillatum, Ambystoma macrodactylum croceum*, and *Ambystoma tigrinum californiense*. *Herpetologica* 24:273–84.

Anderson, J. D., & R. E. Graham. 1967. Vertical migration and stratification of larval *Ambystoma*. *Copeia* 1967:371–4.

Andrewartha, H. G., & L. C. Birch. 1954. *The Distribution and Abundance of Animals*. University of Chicago Press.

Arnold, S. J. 1976. Sexual behavior, sexual interference and sexual defense in the salamanders *Ambystoma maculatum, Ambystoma tigrinum* and *Plethodon jordani*. *Zeitschrift für Tierpsychologie* 42:247–300.

Arnold, S. J., & R. J. Wassersug. 1978. Differential predation on metamorphic anurans by garter snakes (*Thamnophis*): social behavior as a possible defense. *Ecology* 59:1014–22.

Avery, R. A. 1968. Food and feeding relations of three species of *Triturus* (Amphibia Urodela) during the aquatic phases. *Oikos* 19:408–12.

Ayala, F. J., M. E. Gilpin, & J. G. Ehrenfeld. 1973. Competition between species. Theoretical models and experimental tests. *Theoretical Population Biology* 4:331–56.

Bailey, J. R. 1937. Notes on plethodont salamanders of the southeastern United States. *Occasional Papers of the Museum of Zoology, University of Michigan* 364:1–10.

Baltensweiler, W. 1968. The cyclic population dynamics of the grey larch tortrix, *Zeiraphera griseana* Hubner (= *Semiasia diniana* Guenee) (Lepidoptera: Tortricidae). In: *Insect Abundance*, T. R. E. Southwood, ed., pp. 88–97. Oxford, Blackwell Scientific.

Beaver, R. A. 1977. Non-equilibrium "island" communities: Diptera breeding in dead snails. *Journal of Animal Ecology* 46:783–98.

Bell, G. 1974. The reduction of morphological variation in natural populations of smooth newt larvae. *Journal of Animal Ecology* 43:115–28.

– 1977. The life of the smooth newt (*Triturus vulgaris*) after metamorphosis. *Ecological Monographs* 47:279–99.

Bell, G., & J. H. Lawton. 1975. The ecology of the eggs and larvae of the smooth newt [*Triturus vulgaris* (Linn.)]. *Journal of Animal Ecology* 44:393–423.

Bishop, S. C. 1941. *The Salamanders of New York*. Albany, New York State Museum.
Bobka, M. S., R. G. Jaeger, & D. C. McNaught. 1981. Temperature dependent assimilation efficiencies of two species of terrestrial salamanders. *Copeia* 1981:417–21.
Bowman, R. I. 1961. Morphological differentiation and adaptation in the Galapagos finches. *University of California Berkeley Publications in Zoology* 58:1–302.
Branch, L. C., & R. Altig. 1981. Nocturnal stratification of three species of *Ambystoma* larvae. *Copeia* 1981:870–3.
Brandon, R. A., & J. E. Huheey. 1971. Movements and interactions of two species of *Desmognathus* (Amphibia: Plethodontidae). *American Midland Naturalist* 86:86–92.
Brenchley, G. A. 1979. Community matrix models: inconsistent results using Vandermeer's data. *American Naturalist* 113:456–9.
– 1981. A reply to Thomas and Pomerantz with a note on diffuse competition. *American Naturalist* 117:386–9.
Brodie, E. D., Jr. 1968. Investigations on the skin toxin of the red-spotted newt, *Notophthalmus viridescens viridescens*. *American Midland Naturalist* 80:276–80.
Brophy, T. E. 1980. Food habits of sympatric larval *Ambystoma tigrinum* and *Notophthalmus viridescens*. *Journal of Herpetology* 14:1–6.
Brown, J. H., & G. A. Lieberman. 1973. Resource utilization and coexistence of seed-eating desert rodents in sand dune habitats. *Ecology* 54:788–97.
Bruce, R. C. 1971. Life cycle and population structure of the salamander *Stereochilus marginatus* in North Carolina. *Copeia* 1971:234–46.
– 1976. Population structure, life history, and evolution of paedogenesis in the salamander *Eurycea neotenes*. *Copeia* 1976:242–9.
– 1979. Evolution of paedomorphosis in salamanders of the genus *Gyrinophilus*. *Evolution* 33:998–1000.
– 1980. A model of the larval period of the salamander *Gyrinophilus porphyriticus* based on size frequency distributions. *Herpetologica* 36:78–86.
– 1982. Larval periods and metamorphosis in two species of salamanders of the genus *Eurycea*. *Copeia* 1982:117–27.
– 1985. Larval periods, population structure and the effects of stream drift in larvae of the salamanders *Desmognathus quadramaculatus* and *Leurognathus marmoratus* in a southern Appalachian stream. *Copeia* 1985:847–54.
Burton, T. M. 1976. An analysis of the feeding ecology of the salamanders (Amphibia, Urodela) of the Hubbard Brook experimental forest, New Hampshire. *Journal of Herpetology* 10:187–204.
Burton, T. M., & G. E. Likens. 1975a. Salamander populations and biomass in the Hubbard Brook Experimental Forest, New Hampshire. *Copeia* 1975:541–6.
– 1975b. Energy flow and nutrient cycling in salamander populations in the Hubbard Brook Experimental Forest, New Hampshire. *Ecology* 56:1068–80.
Bush, A. O., & J. C. Holmes. 1983. Niche separation and the broken-stick model: use with multiple assemblages. *American Naturalist* 122:849–55.
Carr, D. E., & D. H. Taylor. 1985. Experimental evaluation of population interactions among three sympatric species of *Desmognathus*. *Journal of Herpetology* 19:507–14.

Case, T. J., & M. E. Gilpin. 1974. Interference competition and niche theory. *Proceedings of the National Academy of Sciences USA* 71:3073–7.
Clements, F. E., & V. E. Shelford. 1939. *Bio-ecology*. New York, Wiley.
Cohen, J. E. 1971. Mathematics as metaphor. *Science* 172:674–5.
– 1978. *Food Webs and Niche Space*. Princeton, N.J., Princeton University Press.
Cole, L. C. 1954. The population consequences of life history phenomena. *Quarterly Review of Biology* 29:103–37.
Connell, J. H. 1980. Diversity and the coevolution of competitors, or the ghost of competition past. *Oikos* 35:131–8.
– 1983. On the prevalence and relative importance of interspecific competition: evidence from field experiments. *American Naturalist* 122:661–96.
– 1985. On testing models of competitive coevolution. *Oikos* 45:298–300.
Connell, J. H., & W. P. Sousa. 1983. On the evidence needed to judge ecological stability or persistence. *American Naturalist* 121:789–824.
Connor, E. F., & D. Simberloff. 1978. Species number and compositional similarity of the Galapagos flora and avifauna. *Ecological Monographs* 48:219–48.
– 1979. The assembly of species communities: chance or competition? *Ecology* 60:1132–40.
– 1984. Neutral models of species' coocurrence patterns. In: *Ecological Communities: Conceptual Issues and the Evidence*, D. S. Strong, Jr., D. Simberloff, L. G. Abele, and A. B. Thistle, eds., pp. 316–21, 341–3. Princeton, N.J., Princeton University Press.
Culver, D. C. 1970. Analysis of simple cave communities: niche separation and species packing. *Ecology* 51:949–58.
Curtis, J. T., & R. P. McIntosh. 1951. An upland forest continuum in the prairie-forest border region of Wisconsin. *Ecology* 32:476–96.
Danstedt, R. T. 1975. Local geographic variation in demographic parameters and body size of *Desmognathus fuscus* (Amphibia: Plethodontidae). *Ecology* 56:1054–67.
Darwin, C. S. 1859. *On the Origin of Species by Means of Natural Selection* (6th ed., 1872). London, Murray.
Dawkins, R., & J. R. Krebs. 1979. Arms races between and within species. *Proceedings of the Royal Society of London, Biological Sciences* 205:489–511.
Dayton, P. K. 1973. Two cases of resource partitioning in an intertidal community: making the right prediction for the wrong reason. *American Naturalist* 107:662–70.
DeBach, P., ed. 1964. *Biological Control of Insect Pests and Weeds*. London, Chapman & Hall.
Deevey, E. S., Jr. 1947. Life tables for natural populations of animals. *Quarterly Review of Biology* 22:283–314.
De Jongh, H. J. 1968. Functional morphology of the jaw apparatus of larval and metamorphosing *Rana temporaria* L. *Netherlands Journal of Zoology* 18:1–103.
Diamond, J. M. 1975. Assembly of species communities. In: *Ecology and Evolution of Communities*, M. L. Cody & J. M. Diamond, eds., pp. 342–444. Cambridge, Mass., Harvard University Press.
– 1978. Niche shifts and the rediscovery of interspecific competition. *American Scientist* 66:322–31.
– 1982. Shifts in species incidences and occurrence frequencies due to competition:

musical chairs on islands. *Proceedings of the National Academy of Sciences USA* 79:2420–4.
Diamond, J. M., & M. E. Gilpin. 1982. Examination of the "null" model of Connor and Simberloff for species co-ocurrences on islands. *Oecologia (Berlin)* 52:64–74.
Dice, L. R. 1952. *Natural Communities*. Ann Arbor, University of Michigan Press.
Dodson, S. I. 1970. Complementary feeding niches sustained by size-selective predation. *Limnology and Oceanography* 15:131–7.
Dodson, S. I., & V. E. Dodson. 1971. The diet of *Ambystoma tigrinum* larvae from western Colorado. *Copeia* 1971:614–24.
Dumas, P. C. 1956. The ecological relations of sympatry in *Plethodon dunni* and *Plethodon vehiculum*. *Ecology* 37:485–95.
Dunham, A. E. 1980. An experimental study of interspecific competition between the iguanid lizards *Sceloporus merriami* and *Urosaurus ornatus*. *Ecological Monographs* 50:309–30.
Dunn, E. R. 1917a. The salamanders of the genera *Desmognathus* and *Leurognathus*. *Proceedings of the United States National Museum* 53:393–433.
– 1917b. Reptile and amphibian collections from the North Carolina mountains, with especial reference to salamanders. *Bulletin of the American Museum of Natural History* 37:593–634.
– 1926. *The Salamanders of the Family Plethodontidae*, Northampton, Mass., Smith College Fiftieth Anniversary Publications.
– 1928. The habitats of Plethodontidae. *American Naturalist* 62:236–48.
Efford, I. E., & J. A. Mathias. 1969. A comparison of two salamander populations in Marion Lake, British Columbia. *Copeia* 1969:723–36.
Elton, C. 1946. Competition and the structure of ecological communities. *Journal of Animal Ecology* 15:54–68.
Feldman, M. W., & J. Roughgarden. 1975. A population's stationary distribution and chance of extinction in a stochastic environment with remarks on the theory of species packing. *Theoretical Population Biology* 7:197–207.
Fisher, R. A., A. S. Corbet, & C. B. Williams. 1943. The relation between the number of species and the number of individuals in a random sample of an animal population. *Journal of Animal Ecology* 12:42–58.
Formanowicz, D. R., & E. D. Brodie. 1982. Relative palatabilities of members of a larval amphibian community. *Copeia* 1982:91–7.
Fraser, D. F. 1976a. Empirical evaluation of the hypothesis of food competition in salamanders of the genus *Plethodon*. *Ecology* 57:459–71.
– 1976b. Coexistence of salamanders in the genus *Plethodon*: a variation of the Santa Rosalia theme. *Ecology* 57:238–51.
Frazer, D. 1983. *Reptiles and Amphibians in Britain*. London, William Collins Sons.
Gause, G. F. 1934. *The Struggle for Existence*. Baltimore, Williams & Wilkins.
Gill, D. E. 1972. Intrinsic rates of increase, saturation densities and competitive ability. I. An experiment with *Paramecium*. *American Naturalist* 106:461–71.
– 1974. Intrinsic rate of increase, saturation density, and competitive ability. II. The evolution of competitive ability. *American Naturalist* 108:103–16.
– 1978. The metapopulation ecology of the red-spotted newt, *Notophthalmus viridescens* (Rafinesque). *Ecological Monographs* 48:145–66.

– 1979. Density dependence and homing behavior in adult red-spotted newts *Notophthalmus viridescens* (Rafinesque). *Ecology* 60:800–13.
– 1985. Interpreting breeding patterns from census data: a solution to the Husting dilemma. *Ecology* 66:344–54.
Gill, D. E., K. A. Berven, & B. A. Mock. 1983. The environmental component of evolutionary biology. In: *Population Biology*, C. E. King & P. S. Dawson, eds., pp. 1–36. New York, Columbia University Press.
Gilpin, M. E., & F. J. Ayala. 1973. Global models of growth and competition. *Proceedings of the National Academy of Sciences USA* 70:3590–3.
Gilpin, M. E., & J. M. Diamond. 1984. Are species co-occurrences on islands non-random, and are null hypotheses useful in community ecology? In: *Ecological Communities: Conceptual Issues and the Evidence*, D. R. Strong, Jr., D. Simberloff, L. G. Abele, & A. B. Thistle, eds., pp. 297–315, 332–41. Princeton, N.J., Princeton University Press.
Glass, B. P. 1951. Age at maturity of neotenic *Ambystoma tigrinum mavortium* Baird. *American Midland Naturalist* 84:474–95.
Goh, B. S. 1976. Nonvulnerability of ecosystems in unpredictable environments. *Theoretical Population Biology* 10:83–95.
Goin, C. J., O. B. Goin, & G. R. Zug. 1978. *Introduction to Herpetology*. San Francisco, W. H. Freeman.
Golley, F. B. 1968. Secondary productivity in terrestrial communities. *American Zoologist* 8:53–9.
Gould, S. J. 1977. *Ontogeny and Phylogeny*. Cambridge, Mass., Belknap Press of Harvard University Press.
Grant, P. R. 1968. Bill size, body size, and the ecological adaptations of bird species to competitive situations on islands. *Systematic Zoology* 17:319–33.
– 1981. Speciation and the adaptive radiation of Darwin's finches. *American Scientist* 69:653–63.
– 1983. Inheritance of size and shape in a population of Darwin's finches, *Geospiza conirostris*. *Proceedings of the Royal Society of London, Biological Sciences* 220:219–36.
Grant, P. R., & I. Abbott. 1980. Interspecific competition, island biogeography and null hypotheses. *Evolution* 34:332–41.
Grant, W. C. 1955. Territorialism in two species of salamanders. *Science* 121:137–8.
Griffiths, R. A. 1986. Feeding niche overlap and food selection in smooth and palmate newts, *Triturus vulgaris* and *T. helveticus*, at a pond in Mid-Wales. *Journal of Animal Ecology* 55:201–14.
Hairston, N. G. 1947. The influence of moisture on the local distribution of some North Carolina salamanders. *Bulletin of the Ecological Society of America* 28:54.
– 1949. The local distribution and ecology of the plethodontid salamanders of the southern Appalachians. *Ecological Monographs* 19:47–73.
– 1951. Interspecies competition and its probable influence upon the vertical distribution of Appalachian salamanders of the genus *Plethodon*. *Ecology* 32:266–74.
– 1959. Species abundance and community organization. *Ecology* 40:404–16.
– 1964. Studies on the organization of animal communities. *Journal of Animal Ecology [Supplement]* 33:227–39.

– 1967. Studies on the limitation of a natural population of *Paramecium aurelia*. *Ecology* 48:904–10.
– 1973. Ecology, selection and systematics. *Breviora* 414:1–21.
– 1980a. The experimental test of an analysis of field distributions: competition in terrestrial salamanders. *Ecology* 61:817–26.
– 1980b. Species packing in the salamander genus *Desmognathus*: What are the interspecific interactions involved? *American Naturalist* 115:354–66.
– 1980c. Evolution under interspecific competition. Field experiments on terrestrial salamanders. *Evolution* 34:409–20.
– 1981. An experimental test of a guild: salamander competition. *Ecology* 62:65–72.
– 1983a. Alpha selection in competing salamanders: experimental verification of an a priori hypothesis. *American Naturalist* 122:105–13.
– 1983b. Growth, survival and reproduction of *Plethodon jordani*: trade-offs between selective pressures. *Copeia* 1983:1024–35.
– 1984. Inferences and experimental results in guild structure. In: *Ecological Communities: Conceptual Issues and the Evidence*, D. R. Strong, Jr., D. Simberloff, L. G. Abele, & A. B. Thistle, eds., pp. 19–27. Princeton, N.J., Princeton University Press.
– 1985. The interpretation of experiments on interspecific competition. *American Naturalist* 125:321–5.
– 1986. Species packing in *Desmognathus* salamanders: experimental demonstration of predation and competition. *American Naturalist* 127:266–91.
Hairston, N. G., J. D. Allan, R. K. Colwell, D. J. Futuyma, J. Howell, M. D. Lubin, J. Mathias, & J. H. Vandermeer. 1968. The relationship between species diversity and stability: an experimental approach with protozoa and bacteria. *Ecology* 49:1091–101.
Hairston, N. G., & G. W. Byers. 1954. The soil arthropods of a field in southern Michigan: a study in community ecology. *Contribution from the Laboratory of Vertebrate Biology, University of Michigan* 64:1–37.
Hairston, N. G., K. C. Nishikawa, & S. L. Stenhouse. 1986. Evolution of terrestrial salamanders under competition. Unpublished manuscript.
Hairston, N. G., & C. H. Pope. 1948. Geographic variation and speciation in Appalachian salamanders (*Plethodon jordani* Group). *Evolution* 2:266–78.
Hairston, N. G., F. E. Smith, & L. B. Slobodkin. 1960. Community structure, population control, and competition. *American Naturalist* 94:421–5.
Hall, R. J. 1977. A population analysis of two species of streamside salamanders, genus *Desmognathus*. *Herpetologica* 33:109–13.
Hallett, J. G. 1982. Habitat selection and the community matrix of a desert small-mammal fauna. *Ecology* 63:1400–10.
Hallett, J. G., & S. L. Pimm. 1979. Direct estimate of competition. *American Naturalist* 113:593–600.
Hamilton, W. J. 1932. The food and feeding habits of some eastern salamanders. *Copeia* 1932:83–6.
Harding, J. P. 1949. The use of probability paper for the graphical analysis of polymodal frequency distributions. *Journal of the Marine Biological Association of the United Kingdom* 28:141–53.

Hardisty, M. W. 1961. Studies on an isolated spawning population of the brook lamprey (*Lampetra planeri*). *Journal of Animal Ecology* 30:339–55.

Harris, M. P. 1973. The Galapagos avifauna. *Condor* 75:265–78.

Hassinger, D. D., J. D. Anderson, & G. H. Dalrymple. 1970. The early life history and ecology of *Ambystoma tigrinum* and *Ambystoma opacum* in New Jersey. *American Midland Naturalist* 84:474–95.

Hay, O. P. 1889. Notes on the habits of some ambystomas. *American Naturalist* 23:602–12.

Healy, W. R. 1970. Reduction of neoteny in Massachusetts populations of *Notophthalmus viridescens*. *Copeia* 1970:578–81.

– 1974. Population consequences of alternative life histories in *Notophthalmus viridescens*. *Copeia* 1974:221–9.

Hecht, M. K., & J. L. Edwards. 1976. The determination of parallel or monophyletic relationships: the proteid salamanders – a test case. *American Naturalist* 110:653–77.

Heck, K. L., Jr. 1976. Some critical considerations of the theory of species packing. *Evolutionary Theory* 1:247–58.

Heyer, W. R. 1976. Studies in larval habitat partitioning. *Smithsonian Contributions to Zoology* 242:1–27.

Highton, R. 1962a. Revision of North American salamanders of the genus *Plethodon*. *Bulletin of the Florida State Museum* 6:235–367.

– 1962b. Geographic variation in the life history of the slimy salamander. *Copeia* 1962:597–613.

– 1972. Distributional interactions among eastern North American salamanders of the genus *Plethodon*. In: *The Distributional History of the Biota of the Southern Appalachians*, P. C. Holt, ed., pp. 139–88. Research Division Monograph 4. Blacksburg, Virginia Polytechnic Institute and State University.

Highton, R., & S. A. Henry. 1970. Evolutionary interactions between species of North American salamanders of the genus *Plethodon*. *Evolutionary Biology* 4:211–56.

Highton, R., & A. Larson. 1979. The genetic relationships of the salamanders of the genus *Plethodon*. *Systematic Zoology* 28:579–99.

Highton, R., & R. D. Worthington. 1967. A new salamander of the genus *Plethodon* from Virginia. *Copeia* 1967:617–26.

Holomuzki, J. R. 1980. Synchronous foraging and dietary overlap of three species of plethodontid salamanders. *Herpetologica* 36:109–15.

Horn, H. S. 1966. The measurement of "overlap" in comparative ecological studies. *American Naturalist* 100:419–24.

Houck, L. D. 1977. Life history patterns and reproductive biology of neotropical salamanders. In: *The Reproductive Biology of Amphibians*, D. H. Taylor & S. I. Guttman, eds., pp. 43–72. New York, Plenum Press.

Huffaker, C. B. 1958. Experimental studies on predation: dispersion factors and predator–prey oscillations. *Hilgardia* 27:343–83.

Huheey, J. E., & R. A. Brandon. 1973. Rock-face populations of the mountain salamander, *Desmognathus ochrophaeus*, in North Carolina. *Ecological Monographs* 43:59–77.

Humphrey, R. R. 1922. The multiple testis in urodeles. *Biological Bulletin* 43:45–67.

Hurlbert, S. H. 1970. Predator response to the vermilion-spotted newt (*Notophthalmus viridescens*). *Journal of Herpetology* 4:47–55.

– 1984. Pseudoreplication and the design of ecological field experiments. *Ecological Monographs* 54:187–211.

Husting, E. L. 1965. Survival and breeding structure in a population of *Ambystoma maculatum*. *Copeia* 1965: 352–62.

Hutchinson, G. E. 1959. Homage to Santa Rosalia, or Why are there so many kinds of animals? *American Naturalist* 93:145–59.

– 1978. *An Introduction to Population Ecology*. New Haven, Conn., Yale University Press.

Jaeger, R. G. 1970. Potential extinction through competition between two species of terrestrial salamanders. *Evolution* 24:632–42.

– 1971a. Competitive exclusion as a factor influencing the distribution of two species of terrestrial salamanders. *Ecology* 52:632–7.

– 1971b. Moisture as a factor influencing the distributions of two species of terrestrial salamanders. *Oecologia (Berlin)* 6:191–207.

– 1972. Food as a limited resource in competition between two species of terrestrial salamanders. *Ecology* 53:535–46.

– 1974. Interference or exploitation? A second look at competition between salamanders. *Journal of Herpetology* 8:191–4.

– 1978. Plant climbing by salamanders: periodic availability of plant-dwelling prey. *Copeia* 1978: 686–91.

– 1979. Seasonal spatial distributions of the terrestrial salamander *Plethodon cinereus*. *Herpetologica* 35:90–3.

– 1980. Fluctuations in prey availability and food limitation for a terrestrial salamander. *Oecologia (Berlin)* 44:335–41.

– 1981. Dear enemy recognition and the costs of aggression between salamanders. *American Naturalist* 117:962–74.

Jaeger, R. G., & W. F. Gergits. 1979. Intra- and interspecific communication in salamanders through chemical signals on the substrate. *Animal Behavior* 27:150–6.

Jaeger, R. G., K. C. Nishikawa, & D. E. Barnard. 1983. Foraging tactics of a terrestrial salamander: costs of territorial defense. *Animal Behavior* 31:191–8.

Jasic, F. M. 1981. Abuse and misuse of the term "guild" in ecological studies. *Oikos* 37:397–400.

Jenkins, D., A. Watson, & G. R. Miller. 1967. Population fluctuations in the red grouse *Lagopus lagopus scoticus*. *Journal of Animal Ecology* 36:97–122.

Just, J. J., J. Kraus-Just, & D. A. Check. 1981. A survey of chordate metamorphosis. In: *Metamorphosis: A Problem in Developmental Biology* (2nd ed.), L. I. Gilbert, ed., pp. 265–326. New York, Plenum Press.

Kaplan, D. L. 1977. Exploitative competition in salamanders: test of a hypothesis. *Copeia* 1977:234–8.

Karlin, A. A., S. I. Guttman, & S. L. Rathbun. 1984. Spatial autocorrelation analysis of heterozygosity and geographic distribution in populations of *Desmognathus fuscus* (Amphibia: Plethodontidae). *Copeia* 1984:343–56.

Keen, W. H. 1982. Habitat selection and interspecific competition in two species of plethodontid salamanders. *Ecology* 63:94–102.

– 1985. Habitat selection by two streamside plethodontid salamanders. *Oecologia (Berlin)* 66:437–42.

Keen, W. H., & S. Sharp. 1984. Responses of a plethodontid salamander to conspecific and congeneric intruders. *Animal Behavior* 32:58–65.

Kezer, J. 1952. Thyroxin-induced metamorphosis of the neotenic salamanders *Eurycea tynerensis* and *Eurycea neotenes*. *Copeia* 1952:234–7.

King, C. E. 1964. Relative abundance of species and MacArthur's model. *Ecology* 45:716–27.

King, W. 1936. A new salamander (*Desmognathus*) from the southern Appalachians. *Herpetologica* 1:57–60.

– 1939. A survey of the herpetology of Great Smoky Mountains National Park. *American Midland Naturalist* 21:531–82.

Kleeberger, S. R. 1984. A test of competition in two sympatric populations of desmognathine salamanders. *Ecology* 65:1846–56.

– 1985. Influence of intraspecific density and cover on home range of a plethodontid salamander. *Oecologia (Berlin)* 66:404–10.

Klomp, H. 1966. The dynamics of a field population of the pine looper, *Bupalus piniarius* L. (Lep., Geom.). *Advances in Ecological Research* 3:207–305.

Kneidel, K. A. 1984a. The influence of carcass taxon and size on species composition of carrion-breeding Diptera. *American Midland Naturalist* 111:57–63.

– 1984b. Competition and disturbance in communities of carrion-breeding Diptera. *Journal of Animal Ecology* 53:849–65.

– 1985. Patchiness, aggregation, and the coexistence of competitors for ephemeral resources. *Ecological Entomology* 10:441–8.

Kohn, A. J. 1959. The ecology of *Conus* in Hawaii. *Ecological Monographs* 29:47–90.

– 1971. Diversity, utilization of resources, and adaptive radiation in shallow-water marine invertebrates of tropical oceanic islands. *Limnology and Oceanography* 16:332–48.

Krzysik, A. J. 1979. Resource allocation, coexistence, and the niche structure of a streamside salamander community. *Ecological Monographs* 49:173–94.

Lack, D. 1947. *Darwin's Finches*. Cambridge University Press.

– 1966. *Population Studies of Birds*. Oxford University Press.

Larsen, J. H., & D. J. Guthrie. 1974. Parallelism in the Proteidae reconsidered. *Copeia* 1974:635–43.

Larson, A. 1984. Neontological inferences of evolutionary pattern and process in the salamander family Plethodontidae. *Evolutionary Biology* 17:119–217.

Lawlor, L. R., & J. Maynard Smith. 1976. The coevolution and stability of competing species. *American Naturalist* 110:79–99.

Leslie, P. H., & J. C. Gower. 1960. The properties of a stochastic model for the predator–prey type of interaction between two species. *Biometrika* 47:219–34.

Levin, B. R. 1971. The operation of selection in situations of interspecific competition. *Evolution* 25:249–64.

Levins, R. 1968. *Evolution in Changing Environments. Some Theoretical Explorations*. Princeton, N.J., Princeton University Press.

– 1975. Evolution in communities near equilibrium. In: *Ecology and Evolution of*

Communities, M. L. Cody & J. M. Diamond, eds., pp. 16–50. Cambridge, Mass., Harvard University Press.

Lewontin, R. C. 1965. Selection for colonizing ability. In: *The Genetics of Colonizing Species*, H. G. Baker & G. L. Stebbins, eds., pp. 77–91. New York, Academic Press.

Lindeman, R. L. 1942. The trophic–dynamic aspect of ecology. *Ecology* 23:399–418.

Littleford, R. A., W. F. Keller, & N. E. Phillips. 1947. Studies on the vital limits of water loss in the plethodont salamanders. *Ecology* 28:440–7.

Luckinbill, L. S. 1973. Coexistence in laboratory populations of *Paramecium aurelia* and its predator *Didinium nasutum*. *Ecology* 54:1320–7.

Lynch, J. F. 1985. The feeding ecology of *Aneides flavipunctatus* and sympatric plethodontid salamanders in northwestern California. *Journal of Herpetology* 19:328–52.

MacArthur, R. H. 1955. Fluctuations in animal populations and a measure of community stability. *Ecology* 36:533–6.

– 1957. On the relative abundance of bird species. *Proceedings of the National Academy of Sciences USA* 43:293–5.

– 1958. Population ecology of some warblers of northeastern coniferous forests. *Ecology* 39:599–619.

– 1960. On the relative abundance of species. *American Naturalist* 94:25–36.

– 1966. Note on Mrs. Pielou's comments. *Ecology* 47:1074.

– 1968. The theory of the niche. In: *Population Biology and Evolution*, R. C. Lewontin, ed., pp. 159–76. Syracuse, N.Y., Syracuse University Press.

– 1972. *Geographical Ecology*. New York, Harper & Row.

MacArthur, R. H., & R. Levins. 1964. Competition, habitat selection, and character displacement in a patchy environment. *Proceedings of the National Academy of Sciences USA* 51:1207–10.

– 1967. The limiting similarity, convergence, and divergence of coexisting species. *American Naturalist* 101:377–85.

MacArthur, R. H., & E. O. Wilson. 1963. An equilibrium theory of insular zoogeography. *Evolution* 17:373–87.

– 1967. *The Theory of Island Biogeography*. Princeton, N.J., Princeton University Press.

McGavin, M. 1978. Recognition of conspecific odors by the salamander *Plethodon cinereus*. *Copeia* 1978:356–8.

McMurtrie, R. 1976. On the limit to niche overlap for nonuniform niches. *Theoretical Population Biology* 10: 96–107.

Madison, D. M. 1975. Intraspecific odor preferences between salamanders of the same sex: dependence on season and proximity of residence. *Canadian Journal of Zoology* 53:1356–61.

Maiorana, V. C. 1977. Tail autotomy, functional conflicts and their resolution by a salamander. *Nature* 265:533–5.

– 1978. Difference in diet as an epiphenomenon: space regulates salamanders. *Canadian Journal of Zoology* 56:1017–25.

Martof, B. S. 1962. Some aspects of life history and ecology of the salamander *Leurognathus*. *American Midland Naturalist* 67:1–35.

Martof, B. S., W. M. Palmer, J. R. Bailey, & J. R. Harrison III. 1980. *Amphibians*

and Reptiles of the Carolinas and Virginia. Chapel Hill, University of North Carolina Press.
May, R. M. 1973a. *Stability and Complexity in Model Ecosystems*. Princeton, N.J., Princeton University Press.
– 1973b. Stability in randomly fluctuating versus deterministic environments. *American Naturalist* 107:621–50.
– 1975. Some notes on estimating the competition matrix. *Ecology* 56:737–41.
– ed. 1981a. *Theoretical Ecology. Principles and Applications*. Sunderland, Mass., Sinauer Associates.
– 1981b. Patterns in multi-species communities. In: *Theoretical Ecology. Principles and Applications*, R. M. May, ed., pp. 197–227. Sunderland, Mass., Sinauer Associates.
May, R. M., & R. H. MacArthur. 1972. Niche overlap as a function of environmental variability. *Proceedings of the National Academy of Sciences USA* 69:1109–13.
Means, D. B. 1975. Competitive exclusion along a habitat gradient between two species of salamanders (*Desmognathus*) in western Florida. *Journal of Biogeography* 2:253–63.
Merchant, H. 1972. Estimated population size and home range of the salamanders *Plethodon jordani* and *Plethodon glutinosus*. *Journal of the Washington Academy of Science* 62:248–57.
Moreau, R. E. 1948. Ecological isolation in a rich tropical avifauna. *Journal of Animal Ecology* 17:113–26.
Morin, P. J. 1981. Predatory salamanders reverse the outcome of competition among three species of anuran tadpoles. *Science* 212:1284–6.
– 1983a. Predation, competition, and the composition of larval anuran guilds. *Ecological Monographs* 53: 119–38.
– 1983b. Competitive and predatory interactions in natural and experimental populations of *Notophthalmus viridescens dorsalis* and *Ambystoma tigrinum*. *Copeia* 1983:628–39.
– 1985. Predation intensity, prey survival and injury frequency in an amphibian predator–prey interaction. *Copeia* 1985:638–44.
Morin, P. J., H. M. Wilbur, & R. N. Harris. 1983. Salamander predation and the structure of experimental communities: responses of *Notophthalmus* and microcrustacea. *Ecology* 64:1430–6.
Moss, R., A. Watson, & R. Parr. 1975. Maternal nutrition and breeding success in the red grouse (*Lagopus lagopus scoticus*). *Journal of Animal Ecology* 44:233–44.
Moss, R., A. Watson, & P. Rothery. 1984. Inherent changes in the body size, viability and behavior of a fluctuating red grouse (*Lagopus lagopus scoticus*) population. *Journal of Animal Ecology* 53:171–89.
Munger, J. C., & J. H. Brown. 1981. Competition in desert rodents: an experiment with semipermeable exclosures. *Science* 211:510–12.
Newell, R. L., & G. W. Minshall. 1978. Life history of a multivoltine mayfly, *Tricorythodes minutus*: an example of the effect of temperature on the life cycle. *Annals of the Entomological Society of America* 71:876–81.
Nishikawa, K. C. 1985a. Competition and the evolution of aggressive behavior in two species of terrestrial salamanders. *Evolution* 39:1282–94.

– 1985b. *The Ecology and Evolution of Aggressive Behavior in Two Species of Terrestrial Salamanders.* Ph.D. thesis, University of North Carolina, Chapel Hill.

Noble, G. K. 1927. The plethodontid salamanders: some aspects of their evolution. *American Museum Novitates* 249:1–26.

– 1931. *The Biology of the Amphibia*, New York, McGraw-Hill.

Noble, G. K., & M. K. Brady. 1930. The courtship of the plethodontid salamanders. *Copeia* 1930:52–4.

Nussbaum, R. A. 1976. Geographic variation and systematics of salamanders of the genus *Dicamptodon* Strauch (Ambystomatidae). *Miscellaneous Publications, Museum of Zoology, University of Michigan* 149:1–94.

Organ, J. A. 1958. Courtship and spermatophore of *Plethodon jordani metcalfi*. *Copeia* 1958:251–9.

– 1960. The courtship and spermatophore of the salamander *Plethodon glutinosus*. *Copeia* 1960:34–40.

– 1961. Studies of the local distribution, life history, and population dynamics of the salamander genus *Desmognathus* in Virginia. *Ecological Monographs* 31:189–220.

Orr, L. P., & W. T. Maple 1978. Competition avoidance in salamander larvae of the genus *Desmognathus*. *Copeia* 1978:679–85.

Pacala, S., & J. Roughgarden. 1982. Resource partitioning and interspecific competition in two-species insular *Anolis* lizard communities. *Science* 217:444–6.

– 1985. Population experiments with the *Anolis* lizards of St. Maarten and St. Eustatius. *Ecology* 66:129–41.

Paine, R. T. 1966. Food web complexity and species diversity. *American Naturalist* 100:950–61.

– 1974. Intertidal community structure: experimental studies on the relationship between a dominant competitor and its principal predator. *Oecologia (Berlin)* 15:93–120.

Paine, R. T., & S. A. Levin. 1981. Intertidal landscapes: disturbance and the dynamics of pattern. *Ecological Monographs* 51:145–78.

Patterson, K. K. 1978. Life history aspects of paedogenic populations of the mole salamander, *A. talpoideum*. *Copeia* 1978:649–55.

Pearl, R. 1925. *The Biology of Population Growth*. New York, Alfred A. Knopf.

Petranka, J. W. 1983. Fish predation: a factor affecting the spatial distribution of a stream-breeding salamander. *Copeia* 1983:624–8.

Pianka, E. R. 1969. Sympatry of desert lizards (*Ctenotus*) in western Australia. *Ecology* 50:1012–30.

Pielou, D. P., & E. C. Pielou. 1968. Association among species of infrequent occurrence: the insect and spider fauna of *Polyporus betulinus* (Bulliard) Fries. *Journal of Theoretical Biology* 21:202–16.

Pielou, E. C., & A. N. Arnason. 1966. Correction of one of MacArthur's species-abundance formulas. *Science* 151:592.

Pimentel, D. 1968. Population regulation and genetic feedback. *Science* 159:1432–7.

Pimentel, D., E. H. Feinberg, P. W. Wood, & J. T. Hays. 1965. Selection, spatial distribution, and the coexistence of competing fly species. *American Naturalist* 99:97–109.

Pimm, S. L. 1982. *Food Webs*. New York, Chapman & Hall.

– 1985. Estimating competition coefficients from census data. *Oecologia (Berlin)* 67:588–90.

Pope, C. H. 1928. Some plethodontid salamanders from North Carolina and Kentucky with the description of a new race of *Leurognathus*. *American Museum Novitates* 306:1–19.

Powders, V. N., & W. L. Tietjen. 1974. The comparative food habits of sympatric and allopatric salamanders, *Plethodon glutinosus* and *Plethodon jordani*, in eastern Tennessee and adjacent areas. *Herpetologica* 30:167–75.

Preston, F. W. 1948. The commonness, and rarity, of species. *Ecology* 29:254–83.

– 1962a. The canonical distribution of commonness and rarity: part I. *Ecology* 43:182–215.

– 1962b. The canonical distribution of commonness and rarity: part II. *Ecology* 43:410–32.

Pulliam, H. R., & F. Enders. 1971. The feeding ecology of five sympatric finch species. *Ecology* 52:557–66.

Root, R. B. 1967. The niche exploitation pattern of the blue-gray gnatcatcher. *Ecological Monographs* 37:317–50.

Rosenzweig, M. L., Z. Abramsky, B. Kotler, & W. Mitchell. 1985. Can interaction coefficients be determined from census data? *Oecologia (Berlin)* 66: 194–8.

Roth, V. L. 1981. Constancy in the size ratios of sympatric species. *American Naturalist* 118:394–404.

Roughgarden, J. 1974. Niche width: biogeographic patterns among *Anolis* lizard populations. *American Naturalist* 108:429–42.

– 1976. Resource partitioning among competing species – a coevolutionary approach. *Theoretical Population Biology* 9:388–424.

– 1979. *Theory of Population Genetics and Evolutionary Ecology: An Introduction*. New York, Macmillan.

Roughgarden, J., & M. Feldman. 1975. Species packing and predation pressure. *Ecology* 56:489–92.

Roughgarden, J., D. Heckel, & E. R. Fuentes. 1983. Coevolutionary theory and the biogeography and community structure of *Anolis*. In: *Lizard Ecology: Studies of a Model Organism*, R. B. Huey, E. R. Pianka, & T. W. Schoener, eds., pp. 371–410. Cambridge, Mass., Harvard University Press.

Roughgarden, J., W. P. Porter, & D. G. Heckel. 1981. Resource partitioning of space and its relationship to body temperature in *Anolis* lizard populations. *Oecologia (Berlin)* 50:256–64.

Schmidt, K. P. 1936. Guatemalan salamanders of the genus *Oedipus*. *Zoological Series of Field Museum of Natural History* 20:135–66.

Schoener, T. W. 1965. The evolution of bill size differences among sympatric species of birds. *Evolution* 19:189–213.

– 1974a. Resource partitioning in ecological communities. *Science* 185:27–39.

– 1974b. Some methods for calculating competition coefficients from resource-utilization spectra. *American Naturalist* 108:332–40.

– 1974c. Competition and the form of habitat shift. *Theoretical Population Biology* 6:265–307.

– 1975. Presence and absence of habitat shift in some widespread lizard species. *Ecological Monographs* 45: 233–58.

– 1976. Alternatives to Lotka–Volterra competition: models of intermediate complexity. *Theoretical Population Biology* 10:309–33.
– 1978. Effects of density-restricted food encounter on some single-level competition models. *Theoretical Population Biology* 13:365–81.
– 1982. The controversy over interspecific competition. *American Scientist* 70:586–95.
– 1983. Field experiments on interspecific competition. *American Naturalist* 122:240–85.
– 1985a. On the degree of consistency expected when different methods are used to estimate competition coefficients from census data. *Oecologia (Berlin)* 67: 591–2.
– 1985b. Are lizard population sizes unusually constant through time? *American Naturalist* 126: 633–41.
Selvin, H. C., & A. Stuart. 1966. Data-dredging procedures in survey analysis. *American Statistician* 20:20–3.
Semlitsch, R. D. 1981. Terrestrial activity and summer home range of the mole salamander (*Ambystoma talpoideum*). *Canadian Journal of Zoology* 59:315–22.
– 1983a. Structure and dynamics of two breeding populations of the eastern tiger salamander, *Ambystoma tigrinum*. *Copeia* 1983:608–16.
– 1983b. Burrowing ability and behavior of salamanders of the genus *Ambystoma*. *Canadian Journal of Zoology* 61:616–20.
– 1985a. Reproductive strategy of a facultatively paedomorphic salamander *Ambystoma talpoideum*. *Oecologia (Berlin)* 65:305–13.
– 1985b. Analysis of climatic factors influencing migrations of the salamander *Ambystoma talpoideum*. *Copeia* 1985:477–89.
– 1987. Relationship of pond drying to the reproductive success of the salamander *Ambystoma talpoideum*. *Copeia* 1987:61–9.
Semlitsch, R. D., & J. W. Gibbons. 1985. Phenotypic variation in metamorphosis and paedomorphosis in the salamander *Ambystoma talpoideum*. *Ecology* 66:1123–30.
Sexton, O. J., & J. R. Bizer. 1978. Life history patterns of *Ambystoma tigrinum* in montane Colorado. *American Midland Naturalist* 99:101–18.
Shealey, R. M. 1975. Factors influencing activity in the salamanders *Desmognathus ochrophaeus* and *D. monticola* (Plethodontidae). *Herpetologica* 31:94–102.
Shoop, C. R. 1960. The breeding habits of the mole salamander, *Ambystoma talpoideum* (Holbrook), in southeastern Louisiana. *Tulane Studies in Zoology* 8: 65–82.
– 1974. Yearly variation in larval survival of *Ambystoma maculatum*. *Ecology* 55:440–4.
Simberloff, D. 1984. Properties of coexisting bird species in two archipelagoes. In: *Ecological Communities: Conceptual Issues and the Evidence*, D. S. Strong, Jr., D. Simberloff, L. G. Abele, & A. B. Thistle, eds., pp. 234–53. Princeton, N.J., Princeton University Press.
Simberloff, D., & W. Boecklen. 1981. Santa Rosalia reconsidered: size ratios and competition. *Evolution* 35:1206–28.
Slobodkin, L. B. 1953. An algebra of population growth. *Ecology* 34:513–19.
– 1962. *Growth and Regulation of Animal Populations*. New York, Holt, Rinehart and Winston.

Smith, D. C. 1981. Competitive interactions of the striped plateau lizard (*Sceloporus virgatus*) and the tree lizard (*Urosaurus ornatus*). *Ecology* 62:679–87.

Smith, F. E. 1952. Experimental methods in population dynamics, a critique. *Ecology* 33:441–50.

Snyder, R. C. 1956. Comparative features of the life histories of *Ambystoma gracile* (Baird) from populations at low and high altitudes. *Copeia* 1956: 41–50.

Southerland, M. T. 1985. *Organization in Desmognathine Salamander Communities: The Roles of Habitat and Biotic Interactions*. Ph.D. thesis, University of North Carolina, Chapel Hill.

– 1986a. Behavioral interactions among four species of the salamander genus *Desmognathus*. *Ecology* 67: 175–81.

– 1986b. Behavioral niche expansion in *Desmognathus fuscus* (Amphibia: Caudata: Plethodontidae). *Copeia* 1986:235–7.

– 1986c. Coexistence of three congeneric salamanders: the importance of habitat and body size. *Ecology* 67: 721–8.

– 1986d. The effects of variation in streamside habitats on the composition of mountain salamander communities. *Copeia* 1986:731–41.

Southwood, T. R. E. 1978. *Ecological Methods*. New York, Wiley.

Spiller, D. A. 1986. Consumptive-competition coefficients: an experimental analysis with spiders. *American Naturalist* 127:604–14.

Stenhouse, S. L. 1984. *Coexistence of the Salamanders* Ambystoma maculatum *and* Ambystoma opacum: *Predation and Competition*. Ph.D. thesis, University of North Carolina, Chapel Hill.

– 1985a. Migratory orientation and homing in *Ambystoma maculatum* and *Ambystoma opacum*. *Copeia* 1985:631–7.

– 1985b. Interdemic variation in predation on salamander larvae. *Ecology* 66:1706–17.

Stenhouse, S. L., N. G. Hairston, & A. E. Cobey. 1983. Predation and competition in *Ambystoma* larvae: field and laboratory experiments. *Journal of Herpetology* 17:210–20.

Steward, J. W. 1969. *The Tailed Amphibians of Europe*. New York, Taplinger Publishing.

Stewart, G. D., & E. D. Bellis. 1970. Dispersion patterns of salamanders along a brook. *Copeia* 1970: 86–9.

Stewart, M. M. 1956. The separate effects of food and temperature differences on development of marbled salamander larvae. *Journal of the Elisha Mitchell Scientific Society* 72:47–56.

Strong, D. R., Jr. 1982. Harmonious coexistence of hispine beetles on *Heliconia* in experimental and natural communities. *Ecology* 63:1039–49.

– 1983. Natural variability and the manifold mechanisms of ecological communities. *American Naturalist* 122:636–60.

– 1984. Density-vague ecology and liberal population regulation in insects. In: *A New Ecology: Novel Approaches to Interactive Systems*, P. W. Price, C. N. Slobodchikoff, & W. S. Gaud, eds., pp. 313–37. New York, Wiley.

Sugihara, G. 1980. Minimal community structure: an explanation of species abundance patterns. *American Naturalist* 116:770–87.

Szymura, J. M. 1974. A competitive situation in the larvae of four sympatric species of newts (*Triturus cristatus*, *T. alpestris*, *T. montandoni*, and *T. vulgaris*) living in Poland. *Acta Biologica Cracoviensia* 17:235–52.

Tansley, A. G. 1935. The use and abuse of vegetational concepts and terms. *Ecology* 16:284–307.

Taper, M. L., & T. J. Case. 1985. Quantitative genetic models for the coevolution of character displacement. *Ecology* 66:355–71.

Taylor, E. H. 1944. The genera of plethodont salamanders in Mexico, Part I. *University of Kansas Science Bulletin* 30:189–232.

Terman, C. R. 1968. Population dynamics. In: *Biology of Peromyscus (Rodentia)*, J. A. King, ed., pp. 412–50. Special publication no. 2, American Society of Mammalogists.

Thomas, W. R., & M. J. Pomerantz. 1981. Feasibility and stability in community dynamics. *American Naturalist* 117:381–5.

Thurow, G. 1976. Aggression and competition in eastern *Plethodon* (Amphibia, Urodela, Plethodontidae). *Journal of Herpetology* 10:277–91.

Tilley, S. G. 1968. Size-fecundity relationships and their evolutionary implications in five desmognathine salamanders. *Evolution* 22:806–16.

– 1973a. Life histories and natural selection in populations of the salamander *Desmognathus ochrophaeus*. *Ecology* 54:3–17.

– 1973b. Observations on the larval period and female reproductive ecology of *Desmognathus ochrophaeus* (Amphibia: Plethodontidae) in western North Carolina. *American Midland Naturalist* 89:394–407.

– 1974. Structures and dynamics of populations of the salamander *Desmognathus ochrophaeus* Cope in different habitats. *Ecology* 55:808–17.

– 1977. Studies of life histories and reproduction in North American plethodontid salamanders. In: *The Reproductive Biology of Amphibians*, D. H. Taylor & S. I. Guttman, eds., pp. 1–41. New York, Plenum Press.

– 1980. Life histories and comparative demography of two salamander populations. *Copeia* 1980:806–21.

– 1981. A new species of *Desmognathus* (Amphibia: Caudata: Plethodontidae) from the southern Appalachian mountains. *Occasional Paper of the Museum of Zoology University of Michigan* 695:1–23.

Tilley, S. G., R. B. Merritt, B. Wu, & R. Highton. 1978. Genetic differentiation in salamanders of the *Desmognathus ochrophaeus* complex (Plethodontidae). *Evolution* 32:93–115.

Tinkle, D. W. 1967. The life and demography of the side-blotched lizard, *Uta stansburiana. Miscellaneous Publications, Museum of Zoology, University of Michigan* 132:1–182.

– 1982. Results of experimental density manipulation in an Arizona lizard community. *Ecology* 63:57–65.

Tristram, D. A. 1977. Intraspecific olfactory communication in the terrestrial salamander *Plethodon cinereus*. *Copeia* 1977:597–600.

Turelli, M. 1978. A reexamination of stability in randomly varying versus deterministic environments with comments on the stochastic theory of limiting similarity. *Theoretical Population Biology* 13:244– 67.

Twitty, V. 1966. *Of Scientists and Salamanders*. San Francisco, W. H. Freeman.

Vandermeer, J. H. 1969. The competitive structure of communities: an experimental approach with protozoa. *Ecology* 50:362–71.

– 1981a. *Elementary Mathematical Ecology*. New York, Wiley.
– 1981b. A further note on community models. *American Naturalist* 117:379–80.
Vandermeer, J. H., & R. H. MacArthur. 1966. A reformulation of alternative (b) of the broken stick model of species abundance. *Ecology* 47:139–40.
Varley, G. C. 1949. Population changes in German forest pests. *Journal of Animal Ecology* 18:117–22.
Vial, J. L. 1968. The ecology of the tropical salamander, *Bolitoglossa subpalmata*, in Costa Rica. *Revista de Biologia Tropical* 15:13–115.
Wake, D. B., & J. F. Lynch. 1976. The distribution, ecology, and evolutionary history of plethodontid salamanders in tropical America. *Natural History Museum of Los Angeles County Science Bulletin* 25:1–65.
Walters, B. 1975. Studies of interspecific predation within an amphibian community. *Journal of Herpetology* 9:267–79.
Werner, E. E. 1977. Species packing and niche complementarity in three sunfishes. *American Naturalist* 111:553–78.
Werner, E. E., & D. J. Hall. 1976. Niche shifts in sunfishes: experimental evidence and significance. *Science* 191:404–6.
– 1977. Competition and habitat shift in two sunfishes (Centrarchidae). *Ecology* 58:869–76.
– 1979. Foraging efficiency and habitat switching in competing sunfishes. *Ecology* 60:256–64.
Whitaker, J. O., & D. C. Rubin. 1971. Food habits of *Plethodon jordani metcalfi* and *Plethodon jordani shermani* from North Carolina. *Herpetologica* 27:81–6.
Whitford, W. G., & A. Vinegar. 1966. Homing, survivorship, and overwintering of larvae in spotted salamanders, *Ambystoma maculatum*. *Copeia* 1966:515–19.
Whittaker, R. H. 1952. A study of summer foliage insect communities in the Great Smoky Mountains. *Ecological Monographs* 22:1–44.
– 1965. Dominance and diversity in land plant communities. *Science* 147:250–60.
Wiens, J. A. 1986. Spatial scale and temporal variation in studies of shrubsteppe birds. In: *Community Ecology*, J. Diamond & T. J. Case, eds., pp. 154–72. New York, Harper & Row.
Wiens, J. A., J. T. Rotenberry, & B. Van Horne. 1986. A lesson in the limitations of field experiments: shrubsteppe birds and habitat alteration. *Ecology* 67: 365–76.
Wilbur, H. M. 1971. The ecological relationship of the salamander *Ambystoma laterale* to its all-female, gynogenetic associate. *Evolution* 25:168–79.
– 1972. Competition, predation, and the structure of the *Ambystoma–Rana sylvatica* community. *Ecology* 53:3–21.
Wilbur, H. M., & J. P. Collins. 1973. Ecological aspects of amphibian metamorphosis. *Science* 182:1305–14.
Wilbur, H. M., P. J. Morin, & R. N. Harris. 1983. Salamander predation and the structure of experimental communities: anuran responses. *Ecology* 64:1423–9.
Wilder, I. W. 1924. The relations of growth to metamorphosis in *Eurycea bislineata* (Green). *Journal of Experimental Zoology* 40:1–112.
Williams, C. B. 1947. The generic relations of species in small ecological communities. *Journal of Animal Ecology* 16:11–18.
– 1951. Intra-generic competition as illustrated by Moreau's records of East African bird communities. *Journal of Animal Ecology* 20:246–53.
Wilson, E. O. 1984. *Biophilia*. Cambridge, Mass., Harvard University Press.

Wise, D. H. 1981. A removal experiment with darkling beetles: lack of evidence for interspecific competition. *Ecology* 62:727–38.
Worthington, R. D. 1968. Observations on the relative sizes of three species of salamander larvae in a Maryland pond. *Herpetologica* 24:242–6.
– 1969. Additional observations on sympatric species of salamander larvae in a Maryland pond. *Herpetologica* 25:227–9.
Wrobel, D. J., W. F. Gergits, & R. G. Jaeger. 1980. An experimental study of interference competition among terrestrial salamanders. *Ecology* 61:1034–9.

Name index

Subject index